SpringerBriefs in Applied Sciences and Technology

More information about this series at http://www.springer.com/series/8884

Brajesh Kumar Kaushik
Manoj Kumar Majumder

Carbon Nanotube Based VLSI Interconnects

Analysis and Design

Springer

Brajesh Kumar Kaushik
Manoj Kumar Majumder
Department of Electronics
  and Communication Engineering
Indian Institute of Technology Roorkee
Roorkee
Uttarakhand
India

ISSN 2191-530X   ISSN 2191-5318   (electronic)
ISBN 978-81-322-2046-6   ISBN 978-81-322-2047-3   (eBook)
DOI 10.1007/978-81-322-2047-3

Library of Congress Control Number: 2014946403

Springer New Delhi Heidelberg New York Dordrecht London

Printed on acid-free paper

Springer is part of Springer Science+Business Media (www.springer.com)

Preface

Aggressive scaling of semiconductor process technology over the last several decades has resulted in creation of many new products, such as computers, camera, cell phones, and information appliances. This trend is expected to continue for the coming years and create countless opportunities and challenges. Recent developments in the semiconductor industry shows a rapid increase in interconnect frequency and design complexity. Introduction of newer technologies is now moving toward a 2-year cycle as compared to the traditional 3-year cycle. Though technology scaling helps in addressing design complexity and performance trends, it opens up a whole new spectrum of design validation challenges.

As technology scaling trend continues, interconnect parasitics play dominant role in determining chip performance and functionality. Interconnect delay becomes a significant portion of chip delay and noise/crosstalk caused due to parasitic coupling poses threat to circuit functionality. With increasing demands for high signal speeds coupled with reduced feature sizes, interconnect effects such as signal delay, distortion, and crosstalk become the dominant factors limiting overall performance of high-speed systems. If not considered during the design stage, interconnect effects can cause failed designs.

Advancement of VLSI technology leads to the development of high-speed complex integrated circuits (ICs) in deep submicron and nanoscale regime. Due to shrinking feature size and increasing clock frequency, interconnect plays an important role in determining the overall chip performance. In current scenario, interconnect delay dominates over gate delay. Depending on the length, interconnects can be classified to local, semi-global, and global interconnects. With ever-increasing lengths, global interconnects are prone to large interconnect delays, signal integrity issues, and higher current densities. The global interconnects are widely employed to distribute data, clock, power supply, and ground throughout the entire chip. At global level of interconnects, most of the conventional materials (such as Al or Cu) are susceptible to electromigration due to high current density. This electromigration problem substantially affects the reliability of high-speed circuits. To avoid such problems, researchers are forced to find an alternative solution.

Carbon nanotubes (CNTs) can be considered as an alternative interconnect material for current nanoscale technologies. After discovery in 1991, CNTs have received tremendous research interest for their unique mechanical, electrical, thermal, and chemical properties. The sp^2 bonding in graphene is even stronger than the sp^3 bonds in diamond that gives CNTs very high mechanical strength. The unique electrical and thermal properties are mainly due to movement of electrons in one-dimension (1D). Due to 1D movement, electrons can be scattered only in backward direction. Mean free path (*mfp*) in high quality nanotubes is in the range of micrometers. This is in contrast to a three-dimensional (3D) metallic wire wherein electrons can be backscattered by a series of small angle scatterings, and therefore *mfps* are in the range of a few tens of nanometers. Due to long *mfp*, the ballistic phenomenon can be observed in CNTs that is responsible for its outstanding electrical and thermal behavior. Moreover, an isolated CNT can carry current density in excess of 10^9 A/cm^2, which can enhance the electrical performance as well as eliminate electromigration reliability concerns that plagues current nanoscale Cu interconnects.

This book primarily focuses on the performance analysis of CNT based interconnects in current research scenario. Different CNT structures are modeled on basis of transmission line theory. Performance comparison for different CNT structures illustrates that CNTs are more promising than Cu or other materials used in global VLSI interconnects. The organization of the book chapters is as follows: Chapter 1 provides an overview of current research scenario and basics of interconnects. Unique crystal arrangements and the physical properties of CNTs are described in Chap. 2. Furthermore, this chapter illustrates the production, purification, and applications of CNTs. Chapter 3 provides a brief review of the research work carried out in the area of CNT interconnect modeling. The geometry and equivalent *RLC* parameters for different single and bundled CNTs are also discussed in this chapter. A comparative analysis of crosstalk and delay for different single and bundled CNT structures is carried out in Chap. 4. Finally, Chap. 5 introduces unique mixed CNT bundle structures and their equivalent electrical models.

We would like to express our sincere gratitude to Mr. Arsalan Alam, Mr. Madimatla Satyanarayana Murthy and Mr. Nisarg D. Pandya for their unrestrained help in successful completion of this book. We also acknowledge for the help rendered by all the research scholars and faculty members of Microelectronics and VLSI group, Indian Institute of Technology Roorkee for constructive technical discussions at various stages of preparation of this book.

Brajesh Kumar Kaushik
Manoj Kumar Majumder

Contents

About the Authors

Brajesh Kumar Kaushik is currently working as Associate Professor in Department of Electronics and Communication Engineering, Indian Institute of Technology Roorkee, Roorkee, India. Dr. Kaushik completed his Ph.D. from Indian Institute of Technology Roorkee in 2007. His research interests include Signal Integrity, Propagation Delay, and Power Dissipation of VLSI Interconnects; Low Power VLSI Design; Electronic Design Automation (EDA)-Circuit; Spintronics; Organic Electronics; FinFET Device Circuit Co-Design and Systems-CAD. Dr. Kaushik has authored more than 60 papers in peer-reviewed international journals and over 80 papers in international conferences. He has also authored many book chapters, which include three chapters in Springer books. Dr. Kaushik is Editor-in-Chief of *International Journal of VLSI Design and Communication System (VLSICS)*, AIRCC Publishing Corporation. He also holds the position of Editor of *Microelectronics Journal (MEJ)*, Elsevier Inc.; *Journal of Engineering, Design and Technology (JEDT)*, Emerald Group Publishing Limited; and *Journal of Electrical and Electronics Engineering Research (JEEER)*, Academic Journals.

Manoj Kumar Majumder is currently pursuing his Ph.D. from Indian Institute of Technology Roorkee, Roorkee, India. Prior to this, he completed his B.Tech from Dr. B.C. Roy Engineering College, Durgapur in 2007, and M.Tech from Bengal Engineering and Science University, Shibpur in 2009. He has also worked as a lecturer in Department of Electronics and Communication Engineering at Durgapur Institute of Advanced Technology and Management (DIATM), West Bengal. He has one book chapter and several journal and conference publications to his credit.

Chapter 1
Interconnects

Abstract This chapter briefs about the modeling and classifications of very large scale integration (VLSI) interconnects in deep submicron and nanoscale technology. In current deep submicron (DSM) technology, the clock frequency rapidly increases with shrinking feature sizes. The demand in higher speed and component density of future IC technology increases the resistivity of aluminum (Al) and copper (Cu) interconnects. Therefore, researchers are forced to find an alternative solution for future high-speed global VLSI interconnects. This chapter introduces carbon nanotube, graphene nanoribbon, silicon nanowire, spintronics, and plasmonics as possible replacement of conventional Al or Cu based interconnects. The properties and various advantages of these materials are discussed briefly in this chapter.

Keywords Interconnects · Very large scale integration (VLSI) · Carbon nanotube (CNT) · Graphene nanoribbon (GNR) · Silicon nanowire · Spintronics · Plasmonics · Optical interconnects

1.1 Introduction

In current deep submicron (DSM) technology, feature size continues to shrink whereas the clock frequency continues to increase. This advancement leads the interconnect technology into a new era where it has to face certain challenges such as electromigration, higher resistivity due to surface boundary scattering, skin effect, signal integrity, delay uncertainty, power dissipation, etc. Presently, the devices are much smaller in dimension and faster in comparison to the interconnects. The overall performance of a chip is determined by the interconnect and not the device or gate delay. If the device size and interconnect dimension are scaled down by S, the intrinsic gate delay would scale down by a factor S, while the local interconnect (connecting adjacent devices) delay remains almost same. However, the major problem arises with the global interconnects wherein the delay increases by a factor of S^2. With technology scaling, wires are placed closer to each other with higher aspect ratios. This leads to large coupling capacitance that causes

© The Author(s) 2015
B.K. Kaushik and M.K. Majumder, *Carbon Nanotube Based VLSI Interconnects*, SpringerBriefs in Applied Sciences and Technology, DOI 10.1007/978-81-322-2047-3_1

crosstalk noise and excessive signal delay. The inductive effects are also facing an upward trend due to higher clock frequencies, faster transition (rise/fall) time and longer signal wires. With the scaling of feature size, more and more transistors are accommodated in an integrated chip. To further increase the functionality and subsequently number of transistors, larger number of interconnect levels are required that leads to multi-layer interconnect system. This results in a significant increase of design complexity that requires new design tools to deal with the newer challenges. Due to the above mentioned challenges, the designing of global interconnects is one of the major design concerns for achieving high-performance chips.

As scaling has continued for more than 30 years, it has yielded faster and denser chips with ever-increasing functionality. With scaling down of technology below 100 nm, device performance substantially improves whereas interconnect performance degrades enormously. Traditionally, IC circuit designers considered only device models in circuit simulations whereas interconnect were neglected. But now in scaled DSM technology, interconnects needs much more attention that otherwise can lead to erroneous design and malfunctioning of chip. Interconnect delay dominates the overall circuit delay that limits the performance and packing density of high-functionality chips. Technology scaling forces the interconnect designers to increase the aspect ratio that leads to higher coupling capacitance. The dominating values of coupling capacitance increases crosstalk induced delay between interconnect wires that leads to uncertainty in arrival time of signals. Therefore, an accurate interconnect modeling is essential to ensure the proper functionality and reliable performance of a multi-million transistor VLSI chip (Mezhiba and Friedman 2002, 2004).

Modeling of on-chip interconnect primarily focuses on the generation of 2D/3D interconnect model libraries using the geometry and material characteristics from foundry's electrical design rules. Therefore, it is desirable to observe the overall flow of interconnect modeling as it assures the accuracy of circuit simulations and eventually, the performance and functionality of high-performance ICs (Srivastava et al. 2005a, b, c).

Figure 1.1 presents the overall flow of modeling that starts with the designing of interconnect test structure (ITS). ITS fabricated on silicon, is used to measure, characterize and extract the essential parameters of interconnect geometry and material.

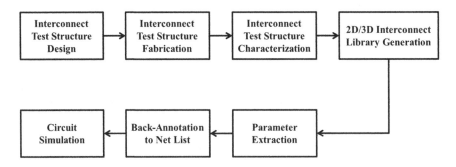

Fig. 1.1 Overall flow of interconnect modeling (Oh et al. 1999)

Based on the characterized interconnect technology parameters (ITP), 2D or 3D interconnect model libraries are generated. Using these interconnect model libraries and device models, essential parameters are extracted from the layout. Later on, they are back-annotated to the netlist and circuit simulations are performed. The accuracy of the circuit simulation depends on the overall accuracy of this flow (Oh et al. 1999).

1.2 Types of Interconnects

In current nanoscale regime, interconnects are categorized in three layers as per their lengths: local, intermediate, and global. The first and second layers of interconnects from the top are global, the third and fourth layers are semi-global/intermediate, and the lowest layers are local. Vias are the metal fillings enabling inter-level wire connections. Interconnects are stacked with dielectric material between two layers or between one layer to transistor, as shown in Fig. 1.2. The wires are separated by interlayer dielectrics (ILD) from level to level and isolated by inter-metal dielectrics (IMD) within the same level (Elgamel and Bayoumi 2003).

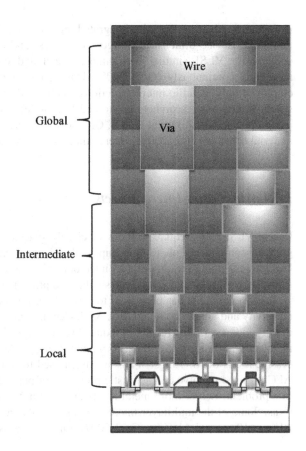

Fig. 1.2 Cross section of stacked interconnects (ITRS 2010)

Detailed descriptions of different interconnect layers are as follows:

(a) *Local*: It consists of very thin lines connecting gates and transistors within an execution unit or a functional block on the chip. Local wires usually span few gates and occupy first and sometimes second metal layers in a multi-level system. The lengths of these wires tend to scale down with technology.

(b) *Semi-global/Intermediate*: It provides clock and signal distribution within a functional block with typical lengths up to 3–4 mm. Intermediate wires are wider and taller than local wires to provide lower resistance signal/clock paths.

(c) *Global*: It provides clock and signal distribution between the functional blocks, and delivers power/ground to all functions on a chip. Global wires, which occupy the top one or two layers, are longer than 4 mm and can be as long as half of the chip perimeter.

1.3 Evolution of Interconnects

Continuous advancements in integrated circuit (IC) technology have resulted in smaller device dimensions, larger chip sizes, and increased complexity. The demand for VLSI circuits with higher speed and component density is also increasing continuously (Goel 2008). During recent past, various IC designers employed only metallic interconnections (such as Al or Cu) but due to their persistent limitations researchers are forced to look after several other possibility of using optical (Kaushik et al. 2007), Graphene (Majumder et al. 2012a) or organic material based (Li et al. 2009a) interconnections in the near future.

1.3.1 Aluminum Interconnects

Recent VLSI chips require millions of closely spaced interconnect lines that integrate the components on a single chip. With the advancement of VLSI technology, it becomes essential to use multilayer interconnections to achieve higher packing densities, lesser transit delays, and with smaller footprint area chips (Goel et al. 2007).

In most cases, Al has been used to form metal interconnects because of its low resistivity and silicon compatibility. As device dimensions are reduced, the current density increases. This results in reduced reliability of VLSI circuits that is due to the electromigration and hillock formation, causing electrical shorts between successive levels of Al. Tungsten has also been used for interconnects and sometimes Al/Cu is used to solve characteristic problems of pure Al. To reduce electromigration problem, two or more metal layers are used in the same level of interconnections. Some of the multilayer interconnect materials are Al/Ti/Cu, Al/Ta/Al, Al/Ni, Al/Cr, Al/Mg, and Al/Ti/Si (Goel 2007). One of the useful methods of reducing the hillocks on silicon-based circuits is to deposit a film of WSi or MoSi between Al and Si substrate. Complete elimination of hillocks was reported when the VLSI interconnections were fabricated by alternatively layering the Al and a refractory metal (Ti or W) layers.

1.3.2   Reason Behind the Replacement of Al by Cu

For several semiconductor technology generations, Al was used as on-chip inter-connects metal and SiO_2 as the inter-level and intra-level insulator. With rapid scaling of feature size to deep submicron levels, the signal delay caused by inter-connects became increasingly significant. This has an adverse effect on reliability of VLSI circuits. The interconnect delay is mostly affected by resistive and capaci-tive parasitics and now inductive parasitics also seem to pose threat. For metallic conductors with electrical resistivity lower than aluminum, the potential options were silver, copper, and gold. Table 1.1 shows the bulk resistivity and thin-film resistivity of these prospective metals.

(a) Gold (Au) has higher resistance to electromigration but shows a little improvement in resistivity. Major demerit is that gold creates deep levels in the bandgap due to diffusion with Si and thus severely affects the electronic properties of a device.

(b) Similarly, silver (Ag) with the lowest resistivity creates deep levels in the sili-con bandgap and diffuses in SiO_2 continuously. Furthermore, building of dif-fusion barrier for Ag is quite difficult and moreover, silver has low resistance to electromigration due to its low melting point.

(c) Copper (Cu) with close to half the resistivity (1.7 $\mu\Omega$ cm) of Al demonstrates ten times better performance in terms of electromigration. Thus, it emerged as the most appropriate material for VLSI interconnect in late 90s.

(d) Adding to the merit, Cu has higher melting point (1,357 K) than aluminum (933 K) that provides an advantage over aluminum in terms of stress and electromigration effects.

The typical VLSI application temperature range ($\approx$373 K) is about 40 percent of the Al melting point and 27.4 % of the Cu melting point (Kaushik et al. 2007). This suggests that mass transport in Cu is generally slower than that of Al at room temperature. Today, Cu is widely used on-chip interconnect material for advanced integrated circuits.

Metal	Bulk resistivity ($\mu\Omega$ cm)	Thin-film resistivity ($\mu\Omega$ cm)
Ag	1.6	–
Cu	1.7	2.1
Au	2.4	4.1
Al	2.65	2.7

Table 1.1   List of conductor materials (Kaushik et al. 2007)

1.3.3 Demerits of Cu Interconnects

In early 2000, it was realized that even Cu is not able to fulfill the demands of high-speed interconnects. With the increasing integration density of the CMOS and higher clock frequency, the requirement of lower resistance and higher bandwidth is the major concern in interconnect design. Therefore, researchers are forced to find an alternative solution to replace Cu due to the following reasons:

(a) The resistivity of Cu interconnects is increasing rapidly under the effects of enhanced grain and surface scattering, longer interconnects, and higher frequency operation.
(b) The resistivity of Cu also increases rapidly due to Joule heating. The increased heating stimulates electromigration induced hillocks and voids.
(c) One key constraint in the conventional scaling of silicon VLSI is the high interconnect related power dissipation per unit area.

Researchers looked aggressively for replacement of Cu interconnect technology since its performance was limited by skin effect, dispersion, signal degradation, power dissipation and electromagnetic interference that actually pronounced at higher frequency range (Kaushik et al. 2007).

1.3.4 Demands in Future Interconnects

The imminent demand in future interconnects are lower parasitics, lower electromigration, less process variability and stress effects at smaller interconnect dimensions. However, at high frequencies, certain problems like skin effect, signal integrity and crosstalk induced propagation delay becomes prominent and difficult to handle. To overcome such problems, VLSI designers need to research with certain materials and their fabrication methods that would be of eminent use in upcoming years. Such advanced materials are carbon nanotubes (CNTs), graphene nanoribbons (GNRs), silicon and metal nanowires, optical interconnects, etc. The physics behind these emerging materials needs to be understood and requires better understanding for their usage as interconnects in nanometer dimensions. The prospects and challenges associated with these material based interconnects are briefly discussed in following sub-sections.

A. Carbon nanotube

Carbon nanotubes (CNTs) have unique atomic arrangement and band structure that is responsible for their outstanding electrical and mechanical properties. These extraordinary properties make CNTs one of the most revered interconnect materials in current nanoscale technology. Some of these properties are briefly discussed below:

(a) They can conduct large current at smaller cross-sectional area without any signal deterioration while simultaneously avoiding electromigration problems that are otherwise prevalent in metallic interconnects.

(b) The resistance of the bundled CNT is about three orders of magnitude lower than single CNT. Thus, it is expected that the CNT bundle would prove to be effective replacement of copper not only for interconnects but for vias also in future VLSI chips (Majumder et al. 2011b).

(c) As the feature size reduces, the performance of Cu interconnects severely degrades due to the increased surface scattering thereby, drastically reducing the effective mean free path. However, in contrast to copper the CNTs supports ballistic flow of electrons with an electron mean free path of several micrometers that strongly motivates researchers to replace Cu by CNTs.

(d) A single-walled CNT (SWNT) results in high contact resistance and characteristic impedance. Therefore, a bundle of closely packed parallel CNTs are preferred (Goel 2008; Majumder et al. 2011a, b). The desired properties of the nanotube bundle includes:

   (i) Low contact resistance with all nanotubes within a bundle.
   (ii) Distance between the nanotubes within the bundle should be as small as possible to have the large nanotube density.
   (iii) Quantum coupling between the nanotubes should be nearly zero.

B. Graphene nanoribbon

Recent development indicates that GNRs have aroused lot of research interests for their potential applications in the area of field effect transistors (FETs) and interconnects (Echtermeyer et al. 2008; Lemme et al. 2007; Ouyang et al. 2006; Gengchiau et al. 2007). Ballistic transport (Gunlycke et al. 2007; Roslyak et al. 2010; Bhattacharya and Mahapatra 2010) in graphene makes it suitable for not only interconnects but for switching transistors also. A monolithic system can be constructed using graphene for transistors and interconnects. The graphene based field effect transistors demonstrate superior mobility than the n or p-type semiconductors based FETs. Moreover, it has been predicted that GNRs will outperform Cu interconnects for widths smaller than 8 nm (Naeemi and Meindl 2007). For nanoscale device dimensions, Cu interconnects are mostly affected by grain boundary and sidewall scatterings (Murali et al. 2009). Therefore, researchers are strongly motivated to find an alternative solution for global VLSI interconnects.

Graphene is a sheet of graphite tightly packed in two-dimensional (2D) honeycomb lattice structures and can be referred as basic building block of graphite, carbon nanotube, graphene nanoribbon, etc. Since GNRs can be considered as unrolled CNTs, most of the electronic properties of GNRs are similar to CNTs. In high quality graphene sheet, the mean free path ranges from 1 to 5 μm (Xu and Srivastava 2009). GNRs can handle current densities more than 10^8 A/cm^2 that is much higher than regular interconnects such as Cu (Li et al. 2009a). It offers higher carrier mobilities that can reach up to 10^5 cm^2 v^{-1} s^{-1} (Li et al. 2009a). For outstanding electrical and thermal properties of GNRs, it is necessary to understand the electronic band structure of graphene. The band structure of graphene is obtained using tight binding approximation (Ragheb and Massoud 2008). As per the tight binding model (Ragheb and Massoud 2008), graphene is a zero bandgap semiconductor or semimetal. Depending on chirality, GNRs can be classified

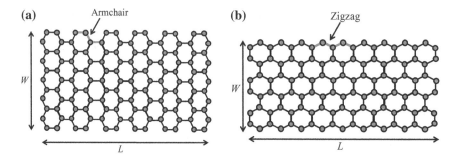

Fig. 1.3 GNR structures for **a** armchair and **b** zigzag chirality

to armchair and zigzag GNRs (ac-GNR and zz-GNR) as shown in Fig. 1.3a, b, respectively. The ac-GNRs can be further categorized to metallic and semiconducting depending on the number of hexagonal rings (N) across the width of GNR that is fixed along with the length. In ac-GNRs, metallic properties depends on the condition of $N = 3p - 1$ or $3p + 2$, whereas $N = 3p$ or $3p + 1$ satisfies the semiconducting properties, where p is any integer. Apart from this, zz-GNRs are always metallic independent of N. Depending on the number of layers formed by the hexagonal rings of carbon atoms, GNRs can be categorized as single-layer GNR (SLGNR) and multi-layer GNR (MLGNR).

C. **Spintronics**

The progress in silicon semiconductor industry during last few decades has largely been supported by the increase in binary information throughput due to the dimensional scaling of components on the chip (Bourianoff 2003). However, dimensional scaling is also accompanied with enormous heat generation. This may limit the extension of Moore's law beyond the technology node meant for year 2024 that corresponds to a minimum feature size of 7.5 nm (Cavin et al. 2006). The post-CMOS devices working with different state variables includes the electron spin, pseudo-spin in graphene, direct and indirect excitons, magnetic domain walls, photons, and plasmons (Galatsis et al. 2009). Among these various new state variables, electron spin is the mostly studied with proven advantages in terms of robustness, nonvolatility and enhanced functionality (Awschalom and Flatte 2007).

The scientific study of spin-electronics is popularly known as Spintronics. In spintronics, the spin of an electron is controlled by an external magnetic field and the polarized electrons as shown in Fig. 1.4. These polarized electrons are used to control the electric current. Spintronics is based on the manipulation of quantum mechanical property of electrons that can be defined as the spin of electrons. Once a spin degree of freedom is added to electrons, it will provide significant versatility and functionality to future electronic products. Magnetic spin properties of electrons are used in various applications such as magnetic memory, magnetic recording (read, write, heads), etc (Zutic et al. 2004; Rakheja and Naeemi 2010).

In digital devices, electrons can store and transmit information in the form of negatively charged subatomic particles. The presence or absence of electrons

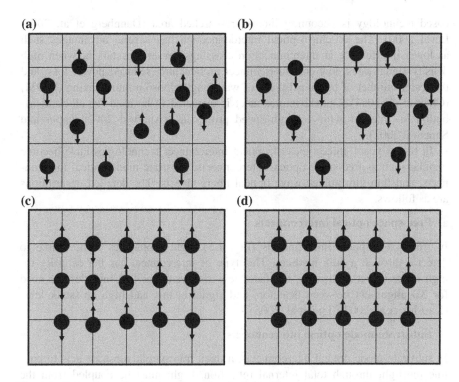

Fig. 1.4 Orientation of spins in spintronics devices with **a** random spin, **b** spin alignment, **c** unmagnetized, and **d** magnetized

within semiconductor or other material represents the zeroes and ones of computer binary code in digital electronic devices. However, in spintronics domain, the information is stored and transmitted using another property of electrons called spin. In simple words, spin is the intrinsic angular momentum of an electron wherein each electron acts like a tiny bar magnet that points either up or down to represent the spin of an electron. Due to the movement of an electron with random spin, the result of net magnetic field effect is zero. External magnetic fields can be applied to store a binary data in the form of ones (all spins up) and zeroes (all spins down), so that the spins are aligned (all up or all down) as per the direction of magnetic field. The effect was first discovered in a device made of alternating magnetic and nonmagnetic layers of electrically conducting materials. The device was known as "spin valve" because when a magnetic field was applied to the device, the spin of its electrons went from all up to all down (Zutic et al. 2004; Rakheja and Naeemi 2012a).

D. **Optical interconnects**

Photons transmitted through fiber are limited only by the dispersion of the medium. Rapid advancements in technology and lowered costs pushes designers to use optical interconnects for on-chip signal transmission. Optical interconnects based technology is becoming intense researched area (Dannberg et al. 2000; Laval 2000). The on-chip optical interconnection has several advantages such as lower interference at interconnection crossings, lesser crosstalk between high density signal lines, low cost interconnect architecture, reduced power loss, and the lesser number of I/O pins by use of wavelength division multiplexing (WDM) (Chen et al. 2005). Optical interconnect is expected to be used for clock distribution in order to achieve high speed processing (Agarwal 2002; Kapur and Saraswat 2002).

In recent years, three types of optical interconnect technologies have become popular such as free-space optical interconnects, substrate-mode optical interconnects, and thin-film guided-wave optical interconnects. The detailed descriptions are as follows:

a. **Free-space optical interconnects**

In free-space optical interconnects (Fig. 1.5), light propagates from source to detector through a bulk medium. This type of interconnect has the capacity for high interconnection density, but faces few disadvantages that includes the need for 3D alignment between detectors and signal beams and high crosstalk level (Kaushik et al. 2007; Tsai et al. 2005).

b. **Substrate-mode optical interconnects**

In substrate-mode optical interconnects, a thick transparent substrate acts to confine the light through total internal reflection. Light must be coupled from the source to the substrate and from the substrate to the detector (Fig. 1.6). Any propagation angle greater than critical is supported, since it is capable of propagating large number of transmission modes. These interconnects that are more robust than free-space optical interconnects require only 2D alignment and demonstrates

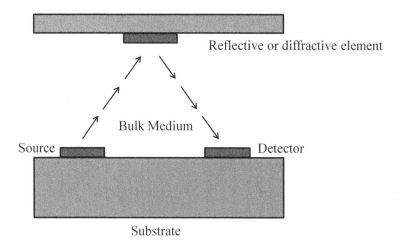

Fig. 1.5 Free-space optical interconnects

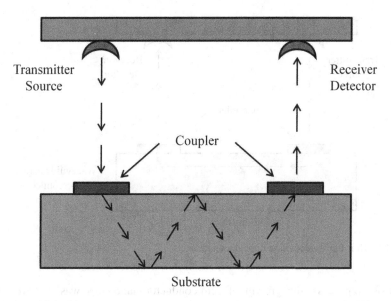

Fig. 1.6 Substrate-mode optical interconnects

lower crosstalk. These optical interconnect cannot compensate for source divergence and requires collimated optical source and further they are bulky and alignment sensitive (Kaushik et al. 2007).

c. **Thin-film guided-wave optical interconnects**

The mechanism of thin-film guided-wave optical interconnects (Fig. 1.7) are also
based on total internal reflection. The primary difference from substrate-mode
optical interconnects is the existence of high-index thin film that is directly above
the substrate and hence the film acts as waveguide. Thin-film guided-wave optical
interconnects, like substrate-mode interconnects, have the advantages of requir-
ing only 2D alignment, are robust, and have low crosstalk (Kaushik et al. 2007;
Tsai et al. 2005). Since, the waveguide is thin; only a discrete, limited number of
modes (and therefore coupling angles) are allowed. It makes coupling of light into
and out of the waveguide difficult. These waveguides are easier to pattern using
photolithographic techniques. They have higher tolerance to source divergence and
allow for the use of integrated optical devices such as direction couplers for routing and splitting the signal.

E. **Silicon nanowires**

Silicon nanowires (SiNWs) are considered as promising material for future
on-chip VLSI interconnects due to their unique physical and electrical properties. In near future, they can be used to link the nanodimensional components
in extremely small circuits. SiNWs can be considered as solid, cylindrical wire
with diameter in the range of few nanometers (~1 nm). Using chemical vapor

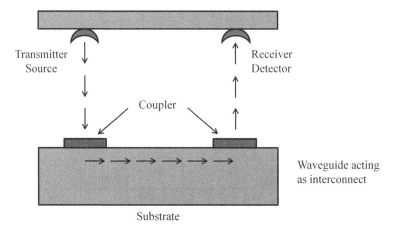

Fig. 1.7   Thin-film guided-wave optical interconnects

deposition process, the growth of semiconductor nanowires was discovered in early 1960s. Such nanowires have been grown by vapor-liquid-solid (VLS) or solution-liquid-solid (SLS). The VLSI growth mechanism is said to be the most versatile method. Under VLSI growth mechanism (Jousseaume and Renard 2010), a metal usually Au nanoparticle (Fig. 1.8a) is used to catalyze NW growth at moderate temperatures (370 °C). The particle appears as small molten ball at the tip of the nanowire as shown in Fig. 1.8b. The nanowires are formed during decomposition of the source gases such as $SiCl_4$ using chemical vapor deposition as shown in Fig. 1.8c. The physical characteristics of nanowires grown in this manner depends on the size and physical properties of the liquid alloy.

$$SiCl_4 + 2H_2 \longrightarrow Si + 4HCl$$

Although, Au is the most convenient metal for nanowire growth, but is a rare earth material, so it is not often used for bulk semiconductor manufacturing process. It is therefore necessary to find some alternative catalysts. Cu is one of the possible solutions but requires high temperatures around 800 °C due to its low eutectic in the Cu/Si system (Jousseaume and Renard 2010; Arbiol et al. 2007). Several physical reasons predict that the conductivity of nanowire is lesser compared to

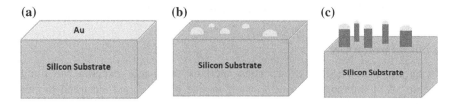

Fig. 1.8   **a** Formation of Au layer, **b** liquid solution of Au, **c** nanowire growth

the corresponding bulk material. One of the major reasons refers to the scattering mechanism near the wire boundaries. This scattering mechanism has significant effect in the smaller wire width dimensions that are below the electron mean free path of the bulk material. However, the scattering effect cannot be observed in CNTs as the motion of electrons in this material falls under the regime of ballistic transport. Therefore, nanowire suffers with high electrical resistance and therefore, finds applications in on-chip resistances and sensor-based designs (Ni et al. 2009).

F. **Plasmonics**

Plasmonics, derived from the greek word 'plasmons', signifies the phenomenon of confinement of light. The frequency of plasmons is equal to the frequency of light beyond the diffraction limit. The packet of energy in the form of 'quanta' is associated with plasmonics due to collective excitation of free electrons in metals. Similarly, oscillation of billions of electrons generates an EM wave (David, 2010). This EM wave can travel along plasmonic materials that generate plasmons at metal and dielectric interface for which a better meta-material with negative refractive index is required. These types of materials such as silver, gold, graphene, etc are known as double negative materials (DNM) (Maffuci et al. 2008a).

Plasmonics is an emerging research area and can be referred as the combination of optics and nanoelectronics. Using the basic theory of plasmonics, a family of novel devices can be produced by confining light with relatively large free-space wavelength to the nanometer scaled devices. Recently, the plasmonic devices faces significant challenges due to the losses encountered in telecommunication and optical frequencies. These large losses seriously limit the practicality of these materials for various novel applications (West et al. 2010).

The operation of plasmonics based device is primarily based on the movement of electrons. At higher frequencies, uniform distribution of electrons cannot be found in metals. Therefore, some of the electrons are crowded at one spot that can be treated as highly negative, while the remaining spots are treated as positive charge (less negative also). As per electromagnetics theory, electric field lines can travel from positive to negative charges and this phenomenon occurs throughout the interconnect wires (Wassel et al. 2012). If some voltage is applied across the terminals of interconnect, these field lines travels along the interface of metal and dielectric with the frequency of light. Depending on the excitations of electrons inside the metal, plasmons are classified as surface plasmon polaritons (SPP) and localized surface plasmons (LSP) (Rakheja and Kumar 2012b). If a surface EM wave propagates in a direction parallel to metal dielectric interface, it is termed as SPP (Rakheja and Kumar 2012) as shown in Fig. 1.9. LSP are collective electron charge oscillations in metallic nanoparticles excited by light.

The materials that are used in metal part of the plasmonic technology are Ag, Au, Al, graphene, etc. Metals like Ag and Au are good for LSP as well as SSP, but graphene is a superior plasmonic material for THz frequency. Graphene is preferable due to its unique band structure, higher carrier mobility, and low loss relative to conventional metal/dielectric interfaces (Maffuci et al. 2008a).

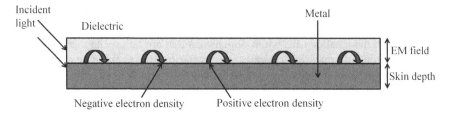

Fig. 1.9 Optical excitations of SPP along a metal dielectric interfaces produces a longitudinal charge density wave with a wave length and skin depth much smaller than the illumination wavelength

1.4 Carbon Nanotubes: The Ultimate Choice

Carbon nanotubes (CNTs) are tiny tubes about 10,000 times thinner than human hair and consist of rolled-up sheets of carbon hexagons. There are mainly two types of CNTs that can have higher structural perfection. Single-walled CNTs (SWNTs) consist of single graphite sheet seamlessly wrapped into a cylindrical tube. Multi-walled CNTs (MWNTs) comprise an array of such nanotubes that are concentrically nested like rings of a tree trunk (Goel 2007).

Electrical transport in metallic SWNTs and MWNTs is ballistic that results in movement of electrons without scattering along the nanotube axis. It enables CNTs a long mean free path in the range of micrometer. Contrastingly, the electrons in Cu can travel only 40–50 nm before scattering. Interestingly, plasmons also propagate easily along the nanotube. Superconductivity has also been observed at low temperatures with transition temperatures of nearly 0.55 K for 1.4 nm diameter SWNTs and nearly 5 K for 0.5 nm SWNTs (Goel 2007).

The unique electrical properties of CNTs such as extremely low electrical resistances are observed due to their unique electronic structure supported by their 1D structure. Primarily, the resistance is observed due to the defects in crystal structures, impurity atoms, or an atom vibrating about its position in the crystal. Due to smaller diameter and higher aspect ratio, the electrons do not scatter much in CNTs that results in relatively lower resistance than Cu. The low resistance ensures that the energy dissipated in CNTs is incredibly small. Thus, the problem of dissipated power density can be properly addressed that otherwise adversely affects the performance of silicon circuits. Current densities of more than 10^{10} A/cm^2 have been observed for the metallic CNTs. Since CNTs do not have any unused bonds, there is no need to grow a film on the surface in order to tie up the free bonds. Moreover, the designer is not restricted to just silicon dioxide as a gate insulator, therefore other superior gate dielectric material can be used that can result in much faster device. Some unique properties of CNTs are listed below:

(a) The carrier transport in CNT is 1D that results in ballistic transport with minimum scattering and lesser power dissipation.

(b) Unlike silicon, all the chemical bonds in carbon atoms are satisfied and thus, there is no need for chemical passivation of free bonds.

(c) The strong C–C covalent bonding is useful for higher mechanical strength, thermal stability, and resistance to electromigration.

(d) The diameter of CNT primarily depends on the rolling up direction of graphene sheets. It is not decided by the fabrication process.

(e) Both active devices and interconnections can be made of semiconducting and metallic nanotubes.

(f) CNTs exhibit higher thermal stability along its axis that is nearly twice as compared to diamond.

All these properties make CNTs the ultimate choice for interconnect material.

Chapter 2
Carbon Nanotube: Properties and Applications

Abstract This chapter presents the unique atomic structure and properties of carbon nanotube (CNT). The electronic band structure of carbon nanotube along with their small size and low dimension are responsible for their unique electrical, mechanical, and thermal properties. This chapter summarizes the electronic band structure of one-dimensional CNTs, various transport properties, and their real-world applications. Additionally, a brief about the production and purification of CNTs is also presented in this chapter.

Keywords Carbon nanotube (CNT) · Single-walled CNT (SWNT) · Double-walled CNT (DWNT) · Multi-walled CNT (MWNT) · Tight binding approximation · Chemical vapor deposition (CVD)

2.1 Introduction

A carbon atom can form various types of allotropes. In 3D structures, diamond and graphite are the allotropes of carbon. Carbon also forms low-dimensional (2D, 1D or 0D) allotropes collectively known as carbon nanomaterials. Examples of such nanomaterials are 1D carbon nanotubes (CNTs) and 0D fullerenes. In the list of carbon nanomaterials, graphene is known as 2D single layer of graphite. The sp^2 bonds in graphene is stronger than sp^3 bonds in diamond that makes graphene the strongest material (Sarkar et al. 2011). The lattice structure of graphene in real space consists of hexagonal arrangement of carbon atoms as shown in Fig. 2.1a. An isolated carbon atoms have four valence electrons in its $2s$, and $2p$ atomic orbitals. While forming into graphene, three atomic orbitals of the carbon atom, $2s$, $2p_x$, and $2p_y$, are hybridized into three sp^2 orbitals. These sp^2 orbitals are in the same plane while the remaining $2p_z$ is perpendicular to other orbitals as shown in Fig. 2.1b (Sarkar et al. 2011). The σ bonds between the adjacent carbon atoms are formed by the sp^2 hybridized orbitals, whereas the $2p_z$ orbitals form the π bonds that are out of the plane of graphene (Javey and Kong, 2009).

© The Author(s) 2015

B.K. Kaushik and M.K. Majumder, *Carbon Nanotube Based VLSI Interconnects*, SpringerBriefs in Applied Sciences and Technology, DOI 10.1007/978-81-322-2047-3_2

(a)   **(b)**

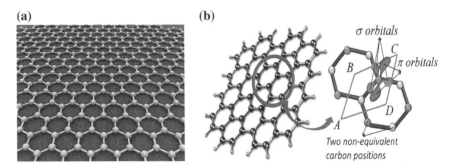

Fig. 2.1 Basic **a** hexagonal and **b** orbital structure of graphene (Reproduced with permission from Sarkar et al. 2011)

Carbon nanotubes (CNTs) are made by rolling up of sheet of graphene into a cylinder. These nanostructures are constructed with length-to-diameter ratio of up to (1.32×10^8):1 (Wang 2009) that is significantly larger than any other material. As their name suggests, the diameter of nanotube is in the order of few nanometers, while they can be up to 18 centimeters in length (Javey and Kong, 2009). CNTs are most promising candidates in the field of nanoelectronics, especially for interconnect applications. Metallic CNTs have aroused a lot of research interest for their applicability as VLSI interconnects due to high thermal stability, high thermal conductivity, and large current carrying capability. A CNT can carry current density in excess of 10^3 MA/cm^2, which can enhance the electrical performance as well as eliminate electromigration reliability concerns that plagues current nanoscale Cu interconnects (Wei et al., 2001). Recent modeling works have revealed that CNT bundle interconnects can potentially offer added advantages over Cu. Moreover, recent experiments have demonstrated that the resistance values as small as 200 Ω can be achieved in CNT bundles.

2.2 Structure and Types of Carbon Nanotubes

To understand the crystal structure of CNTs, it is necessary to understand their atomic structure. Both CNTs and GNRs (graphene nanoribbons) can be understood as structures derived from a graphene sheet, shown in Fig. 2.2. A graphene sheet is a single layer of carbon atoms packed into 2D honeycomb lattice structure. CNT, considered as rolled-up graphene sheet, have the edges of the sheet joint together to form a seamless cylinder. The dashed arrows in Fig. 2.2a, b show the circumferential vector $\vec{C}$, which indicates the rolling up direction for CNT. The vector is defined as $\vec{C} = n_1\hat{a}_1 + n_2\hat{a}_2$ where a_1 and a_2 are the lattice vectors of graphene and n_1 and n_2 are the chiral indices. The chiral indices (n_1, n_2) uniquely defines the chirality, or the rolled-up direction of graphene sheet. Depending on the chiral indices (n_1, n_2), CNTs can be classified to

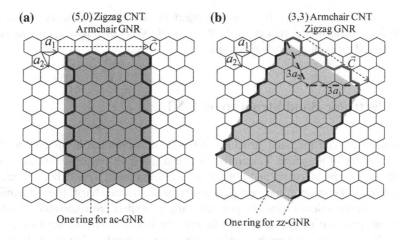

Fig. 2.2 Schematic view of CNT made from graphene sheet **a** zigzag and **b** armchair CNT

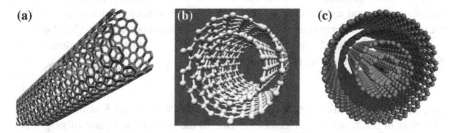

Fig. 2.3 Basic structures of **a** single-walled, **b** double-walled, and **c** multi-walled CNTs

zigzag and armchair structures as shown in Fig. 2.2a, b, respectively. For armchair CNTs, the chiral indices n_1 and n_2 are equal while for zigzag CNTs, n_1 or $n_2 = 0$ (Li et al., 2009b). For other values of indices, CNTs are known as chiral. Depending upon their different structures, CNTs can exhibit metallic or semiconducting properties. By satisfying the condition $n_1 - n_2 = 3i$ (where i is an integer), the armchair CNTs are always metallic, whereas zigzag CNTs are either metallic or semiconducting in nature (Javey and Kong 2009; Li et al. 2009b). Statistically, a natural mix of CNTs will have 1/3rd metallic and 2/3rd semiconducting chiralities.

Depending on the number of concentrically rolled-up graphene sheets, CNTs are also classified to single-walled (SWNT), double-walled (DWNT), and multi-walled CNTs (MWNT) as presented in Fig. 2.3. The structure of SWNT can be conceptualized by wrapping a one-atom-thick layer of graphene into a seamless cylinder (Majumder et al. 2011c). MWNT consists of two or more numbers of rolled-up concentric layers of graphene. DWNT is considered as a special type of MWNT wherein only two concentrically rolled up graphene sheets are present. There are two models to describe the structures of MWNT. In the Russian

Doll model, sheets of graphene are arranged in concentric cylinders, whereas, in the Parchment model (Du et al. 2005), a single sheet of graphene is wrapped around itself resembling a rolled newspaper.

2.3 Electronic Band Structure of CNTs

In order to explain the band structure of CNTs, it is essential to understand the band structure of graphene. Generally, electrical transport properties in graphene are determined by electrons and holes near the Fermi level. This is because electrons near the Fermi level have easy access to the conduction band, leaving behind the holes in the valence band. In graphene, the π orbitals are responsible for the electronic transport properties as they lie near the Fermi level. The band structure of graphene can be obtained by "tight binding approximations" method (Wallace 1947; Minto 2004; Satio et al. 1992). In Fig. 2.4a, a unit cell of graphene is shown with two non-equivalent carbon atoms A and B. With suitable combination of two unit vectors a_1 and a_2, all other atoms can be translated back to either A or B. The reciprocal lattice of graphene with unit vectors, b_1 and b_2 is shown in Fig. 2.4b.

To obtain the band structure of graphene in π orbitals, the solutions of Schrödinger equation is required which states that

$$H\Psi = E\Psi \tag{2.1}$$

where H is the Hamiltonian, Ψ is the total wave function, and E is the energy of electrons in the π orbitals of graphene. Due to the periodic structure of graphene, the total wave function can be constructed from a linear combination of Bloch functions u_i that has a periodicity of the lattice.

In the tight binding approximation, an atomic wave function is used to represent the Bloch function u_i. The u_i for each atom can be constructed from $2p_z$ orbitals of atoms A and B as (Javey and Kong 2009)

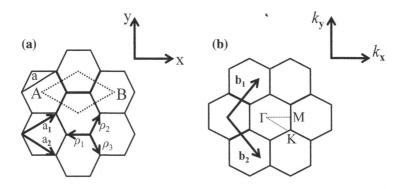

Fig. 2.4 **a** Real and **b** reciprocal space representation of a graphene lattice

$$u_{A(B)} = \frac{1}{\sqrt{N}} \sum_{A(B)} e^{ik.r_{A(B)}} X\left(r - r_{A(B)}\right) \tag{2.2}$$

where $X(r)$ represents the $2p_z$ orbital wave function for an isolated carbon atom. Thus, Ψ in (2.1) can be expressed as (Javey and Kong 2009)

$$\psi = C_A u_A + C_B u_B \tag{2.3}$$

Substituting Eqs. (2.3) into (2.1), the Schrödinger equation can be solved in a matrix form that can be expressed as (Javey and Kong 2009)

$$\begin{pmatrix} H_{AA} & H_{AB} \\ H_{BA} & H_{BB} \end{pmatrix} \begin{pmatrix} C_A \\ C_B \end{pmatrix} = E \begin{pmatrix} S_{AA} & S_{AB} \\ S_{BA} & S_{BB} \end{pmatrix} \begin{pmatrix} C_A \\ C_B \end{pmatrix} \tag{2.4}$$

Here,

$$H_{ij} = \langle u_i | H | u_j \rangle, \quad S_{ij} = \langle u_i | u_j \rangle \tag{2.5}$$

It is easier to neglect the overlap between $2p_z$ wave functions of different atoms, i.e., $S_{AB} = S_{BA} = 0$. For normalized case, the values can be assumed as $S_{AA} = S_{BB} = 1$; then, Eq. (2.4) is simplified to (Javey and Kong 2009)

$$\begin{pmatrix} H_{AA} - E & H_{AB} \\ H_{BA} & H_{BB} - E \end{pmatrix} \begin{pmatrix} C_A \\ C_B \end{pmatrix} = \begin{pmatrix} 0 \\ 0 \end{pmatrix} \tag{2.6}$$

The matrix Eq. (2.6) has a non-trivial solution only when

$$\begin{vmatrix} H_{AA} - E & H_{AB} \\ H_{BA} & H_{BB} - E \end{vmatrix} = 0 \tag{2.7}$$

Further, by symmetry of the graphene lattice (atoms A and B are not distinguishable), it is observed that $H_{AA} = H_{BB}$ and $H_{AB} = H_{BA}^*$. Therefore, Eq. (2.7) leads to the solution

$$E = H_{AA} \mp |H_{AB}| \tag{2.8}$$

H_{AA} ($= H_{BB}$) can be obtained by inserting Eqs. (2.2) into (2.5) as (Javey and Kong 2009)

$$H_{AA} = \frac{1}{N} \sum \sum_{A^*} e^{ik.(r_A - r_A^*)} \int X^*(r - r_A) H X(r - r_A^*) d\tau \tag{2.9}$$

If the effects of the nearest neighbors are considered, the Eq. (2.9) for each atom A(B) with three nearest neighbor B(A) atoms can be obtained as (Javey and Kong 2009)

$$H_{AA} = \int X^*(r - r_A) H X(r - r_A^*) d\tau = E_0 \qquad (2.10)$$

while

$$
\begin{aligned}
H_{AB} &= \frac{1}{N} \sum_A \sum_B e^{ik.(r_A - r_B)} \int X^*(r - r_A) H X(r - r_A^*) d\tau \\
&= \frac{1}{N} \sum_i e^{ik.\rho_i} \int X^*(r) H X(r - \rho_i) d\tau
\end{aligned}
\qquad (2.11)
$$

where ρ_i is a vector connecting atom A to its three nearest neighbor B atoms (as in Fig. 2.4a). By referring to the coordinate system of the graphene in Fig. 2.4a, the following expression can be obtained (Javey and Kong 2009)

$$
\begin{aligned}
H_{AB} &= \left(e^{ik.\rho_1} + e^{ik.\rho_2} + e^{ik.\rho_3} \right) \int X^*(r) H X(r - \rho_1) d\tau \\
&= \gamma_0 \left(e^{-ik_x a / \sqrt{3}} + 2e^{ik_x a / 2\sqrt{3}} \cos\left(\frac{k_y a}{2} \right) \right)
\end{aligned}
\qquad (2.12)
$$

γ_0 represents the strength of exchange interaction between nearest neighbor atoms that is known as the tight-binding integral or transfer integral. Therefore, from Eqs. (2.10) and (2.12), the energy dispersion in Eq. (2.8) can be obtained as (Javey and Kong 2009)

$$E = E_0 \mp \gamma_0 \left(1 + 4\cos\left(\frac{\sqrt{3}k_x a}{2} \right) \cos\left(\frac{k_y a}{2} \right) + 4\cos^2\left(\frac{k_y a}{2} \right) \right)^{1/2} \qquad (2.13)$$

In Eq. (2.13), negative sign represents the valence band of graphene produced by bonding π orbitals, while positive sign denotes conduction band formed by anti-bonding π^* orbitals. The dispersion relation in Eq. (2.13) is plotted in Fig. 2.5 along high-symmetry points in the reciprocal space with $E_0 = 0$. Figure 2.6a, b represent the surface and contour plots of the energy dispersion, respectively. The six K points at the corners of the Brillouin zone are the main feature of the energy dispersion of graphene. At these points, the conduction and valence bands meet resulting in zero bandgap in graphene. It can also be noted that the two K points (K_1 and K_2) are non-equivalent due to symmetry. The circular contour around each K point in Fig. 2.6b indicates the conic shape of dispersion near each K point.

2.3.1 Band Structure of CNTs from Graphene

CNTs can be uniquely defined by chiral vector, $C = n_1 a_1 + n_2 a_2$, where n_1 and n_2 are integers and a_1 and a_2 are unit vectors of the graphene lattice as shown in Fig. 2.7.

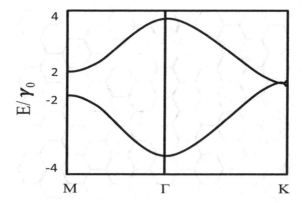

Fig. 2.5 Energy dispersion of graphene along high-symmetry points as indicated in Fig. 2.4b

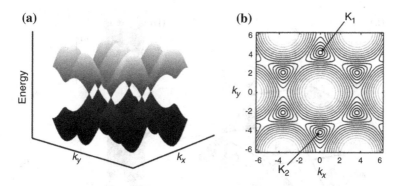

Fig. 2.6 **a** Surface plot and **b** contour plot of the energy dispersion in graphene as given by Eq. (2.13). Note that there are six K points where the bandgap becomes zero. Among the six K points, only two are non-equivalent, denoted by K_1 and K_2 (Reproduced with permission from Javey and Kong 2009)

A graphene sheet is rolled up to form CNT in such a way that two carbon atoms coincide. With wrapping indices n_1 and n_2, CNTs can be uniquely defined and described.

Since CNT is a rolled-up sheet of graphene, an appropriate boundary condition is required to explore the band structure. If CNT can be considered as an infinitely long cylinder, there are two wave vectors associated with it: (1) the wave vector parallel to CNT axis $k_{\parallel}$ that is continuous in nature due to the infinitely long length of CNTs and (2) the perpendicular wave vector $k_{\perp}$ that is along the circumference of CNT. These two wave vectors must satisfy a periodic boundary condition (i.e., the wave function repeats itself as it rotates 2π around a CNT) (Javey and Kong 2009)

$$k_{\perp} \cdot C = \pi D k_{\perp} = 2\pi m \qquad (2.14)$$

where D represents the diameter of CNT and m is an integer. The quantized values of allowed $k_{\perp}$ for CNTs are obtained from the boundary condition.

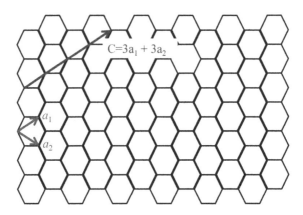

Fig. 2.7 Representation of CNT (single-walled) by a chiral vector, $C = n_1 a_1 + n_2 a_2$

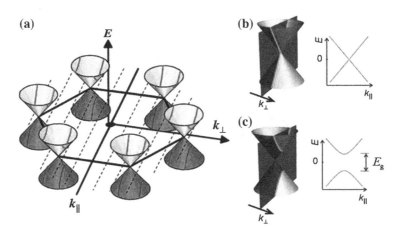

Fig. 2.8 **a** A first Brillouin zone of graphene with conic energy dispersions at six K points. The allowed $k_\perp$ states in CNT are presented by *dashed lines*. The band structure of CNT is obtained by cross sections as indicated. Zoom-ups of the energy dispersion near one of the K points are schematically shown along with the cross sections by allowed $k_\perp$ states and resulting 1D energy dispersions for **b** a metallic CNT and **c** a semiconducting CNT (Reproduced with permission from Javey and Kong 2009)

The cross-sectional cutting of the energy dispersion with the allowed $k_\perp$ states results in 1D band structure of graphene as shown in Fig. 2.8a. This is called zone folding scheme of obtaining the band structure of CNTs. Each cross-sectional cutting gives rise to 1D subband. The spacing between allowed $k_\perp$ states and their angles with respect to the surface Brillouin zone determine the 1D band structures of CNTs. The band structure near the Fermi level is given by allowed $k_\perp$ states that are closest to the K points. When the allowed $k_\perp$ states pass directly through the K points as shown in Fig. 2.8b, the energy dispersion has two linear bands crossing at the Fermi level without a bandgap. However, if the allowed $k_\perp$

states miss the K points as shown in Fig. 2.8c, there are two parabolic 1D bands with an energy bandgap. Therefore, two different kinds of CNTs can be expected depending on the wrapping indices, firstly, the metallic CNTs without bandgap as in Fig. 2.8b, and secondly, the semiconducting CNTs with bandgap as in Fig. 2.8c.

2.3.2 Metallicity and Semiconducting Properties of Zigzag CNTs

Using the approach of 1D subbands discussed in previous subsection, the 1D subband closest to the K points for zigzag CNTs is investigated here. The zigzag CNTs can be either metallic or semiconducting depending on their chiral indices. Since the circumference is na ($C = na_1$), the boundary condition in Eq. (2.14) becomes (Javey and Kong 2009)

$$k_x na = 2\pi m \tag{2.15}$$

There is an allowed k_x that coincides with K point at $(0, 4\pi/3a)$. This condition arises when n has a value in multiple of 3 (n = 3q, where q is an integer). Therefore, by substitution in Eq. (2.15) (Javey and Kong 2009)

$$k_x = \frac{2\pi m}{na} = \frac{3Km}{2n} = \frac{Km}{2q} \tag{2.16}$$

There is always an integer m ($=2q$) that makes k_x pass through K points so that these kinds of CNTs (with $n = 3q$) are always metallic without bandgap as shown in Fig. 2.8b. There are two cases when n is not a multiple of 3. If $n = 3q + 1$, the k_x is closest to the K point at $m = 2q + 1$ (as in Fig. 2.8c).

$$k_x = \frac{2\pi m}{na} = \frac{3Km}{2n} = \frac{3K(2q-1)}{2(3q+1)} = k + \frac{K}{2}\frac{1}{3q+1} \tag{2.17}$$

Similarly, for $n = 3q - 1$, the allowed k_x closest to K is when $m = 2q - 1$; hence (Javey and Kong 2009)

$$k_x = \frac{2\pi m}{na} = \frac{3Km}{2n} = \frac{3K(2q-1)}{2(3q-1)} = k - \frac{K}{2}\frac{1}{3q-1} \tag{2.18}$$

In these two cases, allowed k_x misses the K point by (Javey and Kong 2009)

$$\Delta k_x = \frac{K}{2}\frac{1}{3q\pm 1} = \frac{2}{3}\frac{\pi}{na} = \frac{2}{3}\frac{\pi}{\pi D} = \frac{2}{3D} \tag{2.19}$$

From Eq. (2.19), it is inferred that the smallest misalignment between an allowed k_x and a K point is inversely proportional to the diameter. Thus, from the slope of a cone near K points (Eq. 2.14), the bandgap E_g can be expressed as

$$E_g = 2 \times \left(\frac{\partial E}{\partial k}\right) \times \frac{2}{3D} = 2\hbar v_F \left(\frac{2}{3D}\right) \approx 0.7\text{eV}/D\,(\text{nm}) \tag{2.20}$$

Therefore, semiconducting CNTs ($D = 0.8–3$ nm) exhibit bandgap ranging from 0.2 to 0.9 eV. Depending on the different values of p, three different conditions can be occurred that describes the metallicity and semiconducting properties of CNTs (Javey and Kong 2009):

(a) $p = 0$; metallic with linear subbands crossing at the K points.
(b) $p = 1, 2$; semiconducting with a bandgap, $E_g \sim 0.7$ eV/D (nm).

Similar treatment can also be applied for armchair CNTs (n, n), arriving at the conclusion that they are always metallic.

2.4 Properties of CNTs

The atomic arrangements of carbon atoms are responsible for the unique electrical, thermal, and mechanical properties of CNTs. These properties are discussed below:

2.4.1 Electrical Conductivity

A metallic CNT can be considered as highly conductive material. Chirality, the degree of twist of graphene sheet, determines the conductivity of CNT interconnects. Depending on the chiral indices, CNTs exhibit both metallic or semiconducting properties. The electrical conductivity of MWNTs is quite complex as their inter-wall interactions non-uniformly distribute the current over individual tubes. However, an uniform distribution of current is observed across different parts of metallic SWNT (Shah et al. 2013). Electrodes are placed to measure the conductivity and resistivity of different parts of SWNT rope. The measured resistivity of the SWNT ropes is in the order of 10^{-4} Ω cm at 27 °C, indicating SWNT ropes to be the most conductive carbon fibers (Dresselhaus et al. 2001). It has been reported that an individual SWNT may contain defects that allows the SWNT to behave as a transistor.

2.4.2 Strength and Elasticity

Each carbon atom in a single sheet of graphite is connected via strong chemical bond to three neighboring atoms. Thus, CNTs can exhibit the strongest basal plane elastic modulus and hence are expected to be an ultimate high strength fiber. The elastic modulus of SWNTs is much higher than steel that makes them highly resistant. Although pressing on the tip of nanotube will cause it to bend, the nanotube returns to its original state as soon as the force is removed. This property makes CNTs extremely useful as probe tips for high resolution scanning probe microscopy.

Although, the current Young's modulus of SWNT is about 1 TPa, but a much higher value of 1.8 TPa has also been reported (Hsieh et al. 2006). For different experimental measurement techniques, the values of Young's modulus vary in the range of 1.22 TPa–1.26 TPa depending on the size and chirality of the SWNTs (Dresselhaus et al. 2001). It has been observed that the elastic modulus of MWNTs is not strongly dependent on the diameter. Primarily, the moduli of MWNTs are correlated to the amount of disorder in the nanotube walls (Forro et al. 2002).

2.4.3 Thermal Conductivity and Expansion

CNTs can exhibit superconductivity below 20 K (approximately −253 °C) due to the strong in-plane C–C bonds of graphene. The strong C-C bond provides the exceptional strength and stiffness against axial strains. Moreover, the larger inter-plane and zero in-plane thermal expansion of SWNTs results in high flexibility against non-axial strains.

Due to their high thermal conductivity and large in-plane expansion, CNTs exhibit exciting prospects in nanoscale molecular electronics, sensing and actuating devices, reinforcing additive fibers in functional composite materials, etc. Recent experimental measurements suggest that the CNT embedded matrices are stronger in comparison to bare polymer matrices (Wei et al. 2002). Therefore, it is expected that the nanotube may also significantly improve the thermo-mechanical and the thermal properties of the composite materials.

2.4.4 Field Emission

Under the application of strong electric field, tunneling of electrons from metal tip to vacuum results in field emission phenomenon. Field emission results from the high aspect ratio and small diameter of CNTs. The field emitters are suitable for the application in flat-panel displays. For MWNTs, the field emission properties occur due to the emission of electrons and light. Without applied potential, the luminescence and light emission occurs through the electron field emission and visible part of the spectrum, respectively.

2.4.5 Aspect Ratio

One of the exciting properties of CNTs is the high aspect ratio, inferring that a lower CNT load is required compared to other conductive additives to achieve similar electrical conductivity. The high aspect ratio of CNTs possesses unique electrical conductivity in comparison to the conventional additive materials such as chopped carbon fiber, carbon black, or stainless steel fiber.

2.4.6 Absorbent

Carbon nanotubes and CNT composites have been emerging as perspective absorbing materials due to their light weight, larger flexibility, high mechanical strength and superior electrical properties. Therefore, CNTs emerge out as ideal candidate for use in gas, air and water filtration. The absorption frequency range of SWNT-polyurethane composites broaden from 6.4–8.2 (1.8 GHz) to 7.5–10.1 (2.6 GHz) and to 12.0–15.1 GHz (3.1 GHz) (Wang et al. 2013). A lot of research has already been carried out for replacing the activated charcoal with CNTs for certain ultrahigh purity applications.

2.5 Production of CNTs

A number of methods are used to produce CNTs and fullerenes. In earlier days, fullerenes were produced by vaporizing graphite with a short-pulse, high-power laser method, whereas carbon combustion and vapor deposition processes were used to produce CNTs. Using the earlier method (plasma arcing) of producing CNTs and fullerenes in reasonable quantities, an electric current was applied across two carbonaceous electrodes in an inert gas atmosphere. The plasma arcing method is primarily used to produce fullerenes and CNTs from different carbonaceous materials such as graphite. The CNTs are deposited on the electrode, while the fullerenes appear in the soot. Plasma arcing method can also be applied in the presence of cobalt with 3 % or greater concentration. Different methods for production of CNTs and fullerenes are discussed below:

2.5.1 Arc Discharge Method

The most common and easiest way to produce CNTs is the carbon arc discharge method. Using this method, a complex mixture of components is produced that separates the CNTs from soot and the catalytic metals. Arc vaporization is used to produce CNTs by placing two carbon rods from end to end, separated by an approximate distance of 1 mm, placed in an enclosure. At low pressure, the enclosure is usually filled with inert gas. A high temperature discharge between two electrodes is created by applying a direct current of 50 to 100 A. The surface of the carbon electrodes is vaporized by the high temperature discharge and finally forms a small rod-shaped electrode. The production of CNTs in high yield primarily depends on the uniformity of the plasma arc. The arc discharge method with liquid nitrogen can also be used to produce CNTs.

2.5.2 Laser Method

In 1996, a dual-pulsed laser was used in synthesization techniques for producing CNTs with 70 % purity. Presently the laser vaporization process is used for producing CNTs. In this process, a graphite rod with 50:50 catalyst mixtures of cobalt and nickel at 1200 °C in flowing argon is used to prepare the sample (Ahmed 2010). The uniform vaporization of the target can be achieved by using the initial laser vaporization pulse followed by a second pulse. The amount of deposition of carbon as soot is primarily minimized by the usage of these two successive laser pulses. The larger particles are broken by applying the second laser pulse. The CNTs produced through this process are 10–20 nm in diameter and 100 μm or more in length. The average nanotube diameter and size distribution can vary for different growth temperature, catalyst composition, and other process parameters. In recent years, the arc discharge and laser vaporization methods are used to obtain high quality CNTs in small quantity. However, both methods suffer from the following two drawbacks: (1) The methods use the evaporation of carbon source that follows an unclear approach to scale up the production with industrial standard and (2) the CNTs produced by vaporization method get mixed with residues of carbon. Therefore, it is quite difficult to purify, manipulate, and assemble CNTs for building nanotube device architectures for practical applications.

2.5.3 Chemical Vapor Deposition

From last twenty years, carbon fibers and filaments are produced using chemical vapor deposition of hydrocarbons along with a metal catalyst. In this process, large amount of CNTs is produced by catalytic CVD of acetylene over cobalt and iron. Using the carbon/zeolite catalyst, fullerenes and bundles of SWNTs can be produced along with the MWNTs. Lot of research works have been carried out for the formation of SWNTs/MWNTs from ethylene that is supported by the catalysts such as iron, cobalt, and nickel. The recent research works have also demonstrated the production of SWNTs and DWNTs on molybdenum and molybdenum–iron alloy catalysts. A thin alumina template with or without nickel catalyst is achieved using the CVD of carbon within the pores. Ethylene can be used with reaction temperatures of 545 °C for Nickel-catalyzed CVD and 900 °C for an uncatalyzed process that produces carbon nanostructures with open ends (Che et al. 1998). Methane can also be used as carbon source for synthesization. Catalytic decomposition of H_2/CH_4 mixture over cobalt, nickel, and iron is used to obtain high yields of SWNTs at 1,000 °C. The usage of H_2/CH_4 atmosphere between a non-reducible oxide such as Al_2O_3 or $MgAl_2O_4$ and one or more transition metal oxides can produce the composite powders containing well-dispersed CNTs. Thus, higher proportions of SWNTs and lower proportions of MWNTs can be achieved using the decomposition of CH_4 over the freshly formed nanoparticles.

2.5.4 Ball Milling

One of the simple methods for the production of CNTs is through ball milling followed by subsequent annealing. Thermal annealing is used to produce CNTs from
carbon and boron nitride powder. In this method, graphite powder is placed in a
stainless steel container consisting of four hardened steel balls. The steel container
is purged, and argon is introduced for milling process, at room temperature, for up
to 150 h (Wilson et al. 2002). Using the milling process, the annealing of graphite
powder is carried out under an inert gas flow at 1400 °C for 6 h. This method produces more MWNTs and few SWNTs.

2.5.5 Other Methods

Apart from the above mentioned synthesis methods, CNTs can also be produced using
electrolysis, flame synthesis, synthesis from bulk polymer, use of solar energy and
low-temperature solid pyrolysis. Using the electrolysis method, CNTs are produced
by passing an electric current in molten ionic salt between graphite electrodes. High-
purity carbon rod is used as cathode. At high melting point, the cathode is consumed
and a wide range of nanomaterials are produced. In the flame synthesis method, a portion of the hydrocarbon gas is provided at an elevated temperature that is required for
the hydrocarbon reagents to function. The CNTs can also be synthesized chemically by
using polymers consisting of carbon. This process can produce MWNTs with diameters ranging from 5 nm to 20 nm and a length upto 1 μm (Journet and Bernier, 1998).

Several research groups (Chibante et al. 1993; Laplaze et al. 1996) demonstrated the production of fullerenes using highly concentrated sunlight from a solar
furnace. Using this method, sunlight is focused on a graphite sample to condense
the carbon soot in a cold dark zone of the reactor. The pyrolysis method is used to
synthesize CNTs from metastable carbon containing compounds at relatively low
temperature of 1200 °C to 1900 °C. This method produces MWNTs of diameters
ranging from 10 nm to 25 nm and lengths of 0.1–1 μm.

2.6 Purification of CNTs

After synthesization, purification of CNTs is done by separating it from other enti-
ties, such as amorphous carbon, carbon nanoparticles, residual catalyst and other
unwanted species. The classic chemical techniques are proved as ineffective for
removing the undesirable impurities from CNTs. Therefore, three basic method-
ologies such as gas phase, liquid phase, and intercalation are used to purify the
CNTs from the undesirable impurities.

Generally, a microfiltration operation is performed to remove the nanoparticles
using membrane filters. This methodology removes the amorphous carbon and
unwanted nanoparticles simultaneously without chemically modifying the CNTs.

Furthermore, the smaller chemical substances can be removed by applying only 2–3 mol. nitric acid (Wilson et al. 2002). Using the extended sonication in concentrated acid mixtures, CNTs are cut into smaller segments, forming colloidal suspension in solvents. The solvents can be deposited on substrates or in solution that provides different functional groups attached to the ends and sides of the CNTs. The details of different purification methods are discussed below:

2.6.1 Gas Phase

Thomas Ebbesen and his coworkers developed the most successful technique for purification of nanotubes that is formally known as gas-phase method (Ebbesen 1996). Using this method, the researchers realized that the oxidization process is easier for the nanoparticles with defect as compared to the relatively perfect nanotubes. This method produces significant enrichment of nanotubes.

Recently, NASA Glenn Research Center introduced a new gas-phase method for purification of SWNTs in gram-scale quantities (Glenn Research Center 2002). This purification method uses the combination of high temperature oxidation and repeated extraction with nitric and hydrochloric acid. Using this improved procedure, the stability of the nanotube is significantly improved with a negligible reduction of impurities such as residual catalyst and non-nanotube forms of carbon.

2.6.2 Liquid Phase

Another effective purification method is the liquid-phase methodology. The current liquid-phase purification method follows the essential steps described below:

- Preliminary filtration—used to remove large graphite particles
- Dissolution—used to remove fullerenes and catalyst particles
- Microfiltration
- Centrifugal separation and
- Chromatography

In this process, CNTs are subjected to the liquid phase oxidation step with hydrogen peroxide (H_2O_2) solution. It removes the amorphous carbon without damaging the tube walls. Thus, it becomes quite easy to separate CNTs in the last stage of separation.

2.6.3 Intercalation

Intercalation technique is used to insert a molecule (or ion) into compounds with layered structures. This technique expands the Vander Waals gap between adjacent layers, which requires energy. This energy is usually supplied by the charge transfer between the dopants and the solid.

A research group introduced the intercalation methodology with a variety of materials to purify the nanotubes (Harris 1999). Using the intercalation with Cu chloride, it is preferentially easy to oxidize the nanoparticles away. Thus, this process became popular for purification of nanotubes. In the first stage, the cathode is immersed in a molten copper chloride and potassium chloride mixture for one week at 400 °C. To remove the excess copper chloride and potassium chloride, ion exchanged water is used to wash the mixture of graphitic fragments and the inter-calated nanoparticles. The washed product is slowly heated at 500 °C in a mixture of helium and hydrogen for one hour to reduce the intercalated copper chloride and potassium chloride metal. Finally, the oxidation of the material is carried out, in flowing air, at a rate of 10 °C/min to a temperature of 555 °C. This process provides a fresh sample of cathodic soot that can be treated as fresh nanotube. The drawbacks of this method are: (1) loosing of some amount of nanotubes at oxidation stage and (2) contamination of the final material with residues of intercalates.

2.7 Application of CNTs

CNTs have not only unique atomic arrangements but also have unique properties (Li et al. 2009a) that include large current carrying capability (Wei et al. 2001), long ballistic transport length (Javey and Kong 2009), high thermal conductivity (Collins et al. 2001a), and mechanical strength (Berber et al. 2000). These extraordinary properties of CNTs qualifies them exciting prospects and variety of applications in the area of microelectronics/nanoelectronics (Avorious et al. 2007), spintronics (Tsukagoshi et al. 1999), optics (Misewich et al. 2003), material science (Wang et al. 2000), mechanical (Yu et al. 2000), and biological fields (Yu et al. 2000; Lu et al. 2009). Particularly, in the area of nanoelectronics, CNTs and graphene nanoribbons (GNRs) demonstrates wide range of applications such as energy storage [supercapacitor (Du et al. 2005)] devices; energy conversion devices that includes thermoelectric (Wei et al. 2009) and photovoltaic (Ago et al. 1999) devices; field emission displays and radiation sources (Choi et al. 1999); nanometer semiconductor transistors (Collins et al. 2001a), nanoelectromechanical systems (NEMS) (Dadgour et al. 2008), electrostatic discharge (ESD) protection (*Hyperion Catalysis*), interconnects (Kreupl et al. 2002; Li et al. 2009a), and passives (Li and Banerjee 2008a). The applications of CNTs in different fields are listed below.

2.7.1 Structural

CNTs possesses remarkable properties and qualities as structural materials. Their potential applications include (Jorio et al. 2008):

(a) *Textiles*—CNTs can produce waterproof and tear-resistant fabrics.
(b) *Body armor*—CNT fibers are being used as combat jackets. The jackets are used to monitor the condition of the wearer and to provide protection from bullets.

(c) *Concrete*—CNTs in concrete increases its tensile strength and halt crack propagation.
(d) *Polyethylene*—CNT fibers can be used as polyethylene. The CNT based polyethylene can increase the elastic modulus of the polymers by 30 %.
(e) *Sports equipment*—Golf balls, golf clubs, stronger and lighter tennis rackets, bicycle parts, and baseball bats.
(f) *Bridges*—CNTs may be able to replace steel in suspension and bridges.
(g) *Flywheels*—The high strength/weight ratios of CNTs enable very high rotational speeds.
(h) *Fire protection*—Thin layers of buckypaper can potentially protect the object from fire. The dense, compact layer of CNT or carbon fibers in the form of buckypaper can efficiently reflect the heat.

2.7.2 Electromagnetic

CNTs can be fabricated as electrical conductors, semiconductors and insulators. Such applications include:

(a) *Buckypaper*—Thin nanotube sheets are 250 times stronger and 10 times lighter than steel. They can be used as heat sink for chipboards, backlight for LCD screens, or Faraday cage to protect electrical devices/aeroplanes (Ji et al. 2006). A fabricated device composed of *p*-type and *n*-type elements is shown in Fig. 2.9 (Hu et al. 2010).
(b) *Light bulb filament*—CNTs can be used as alternative to tungsten filaments in incandescent lamps (Jornet and Akyildiz 2010).
(c) *Magnets*—A strong magnetic field can be generated using multi-walled CNTs coated with magnetite (Jornet and Akyildiz 2010).
(d) *Solar cells*—Germanium CNT diode exploits the photovoltaic effect. In some solar cells, nanotubes are used to replace the ITO (indium tin-oxide) to allow the light to pass to the active layers and generate photocurrent (Laplazeb et al. 1997).
(e) *Electromagnetic antenna*—CNTs can act as an antenna for radio and other electromagnetic devices due to its durability, light weight and conductive properties (Jornet and Akyildiz 2010). The skin effect in CNTs is negligible at high frequencies due to additional kinetic inductance. This results in low power dissipation, resulting in high antenna efficiency (Huang et al. 2008).

2.7.3 Electroacoustic

The application of CNT in the field of electroacoustic is:
Loudspeaker—Loudspeakers can be manufactured from sheets of parallel CNTs. Such a loudspeaker can generate sound similar to the sound of lightening producing thunder (Jorio et al. 2008; Avouris et al. 2003, 2007).

Fig. 2.9 A fabricated demo-
device composed of 50 pairs
of *p*-type and *n*-type elements
(Reproduced with permission
from Hu et al. 2010)

2.7.4 Chemical

CNTs finds tremendous applications in the chemical field also, few of them are as
follows:

(a) *Air pollution filter*—CNTs are one of the best materials for air filters because
    they possess high adsorption capacity and large specific area. The conductance
    of CNTs changes when polluted gas comes in its contact. This helps in detect-
    ing and filtering the polluted air (Ong et al. 2010). CNT membranes can suc-
    cessfully filter carbon dioxide from power plant emissions (Lin et al. 2008).
(b) *Water filter*—CNT membranes can aid in filtration. It can reduce distillation
    costs by 75 %. These tubes are so thin that small particles (like water mol-
    ecules) can pass through them, while blocking larger particles (such as the
    chloride ions in salt) (Jorio et al. 2008; Lin et al. 2008).
    CNTs have high active site and controlled distribution of pore size on their
    surface. This increases not only its sorption capabilities, but also its sorption
    efficiency. CNTs have effective sorption capacity over broad pH range (par-
    ticularly for 7 to 10 pH) (Ong et al. 2010).
(c) *Chemical Nanowires*—The CNTs finds their applications in nanowire manufactur-
    ing using materials such as gold, zinc oxide, gallium arsenide, etc. The gold based
    CNT nanowires are very selective and sensitive to hydrogen sulphide (H_2S) detec-
    tion. The zinc oxide (ZnO) based CNT nanowires can be used in applications for
    light emitting devices and harvesters of vibrational energy (Ok et al. 2010).
(d) *Sensors*—CNT based sensors can detect temperature, air pressure, chemical
    gases (such as carbon monoxide, ammonia), molecular pressure, strain, etc.
    The operation of a CNT based sensor is primarily dependent on the genera-
    tion of current/voltage. The electric current is generated by the flow of free
    charged carrier induced in any material. This charge is typically modulated by
    the adsorption of a target on the CNT surface (Martin et al. 2012). A CNT
    based fabricated gas sensing device is shown in Fig. 2.10 (Lin et al. 2013).

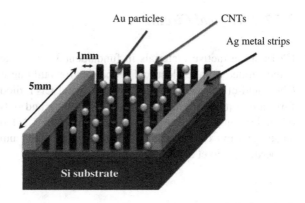

Fig. 2.10 Fabricated gas sensing device (Reproduced with permission from Lin et al. 2013)

2.7.5 Mechanical

The potential application of CNTs can be found in the following areas of mechanical engineering as well :

(a) *Oscillator*—Oscillators based on CNTs have achieved higher speeds than other technologies (>50 GHz) (Jorio et al. 2008). Researchers reported a molecular oscillator with frequencies upto several gigahertz. The operation of this oscillator is primarily based on the low friction and low wear bearing properties of a multi-walled CNT with a diameter ranging from 1 nm to a few tens of nanometers.

(b) *Waterproof*—CNTs can be used to prepare superhydrophobic cotton fabric by dip-coating approach. This approach is solely based on the chemical reactions caused by UV-activated nitrene solution. The solution is used to transform the cotton fabric surface from hydrophilic to superhydrophobic with an apparent water contact angle of 154°. Since CNTs are covalently attached on the surface of the cotton fabric, the superhydrophobicity possesses high stability and chemical durability (Jorio et al. 2008).

2.7.6 Optical

Carbon nanotubes are grown like a field of grass, where each nanotube is separated like a blade of grass. Thus, a particle of light bounces between the nanotubes. In this process, light is completely absorbed and it is converted to heat. Therefore, the absorbance of CNT is extremely high in wide ranges from FUV to FIR (Mizuno et al. 2009) [FUV (Far Ultraviolet): 100-200 nm; FIR (Far Infrared): 50-1000 μm].

2.7.7 Electrical Circuits

CNTs are attractive materials in fundamental science and technology. They have demonstrated unique electrical properties for building electronic devices, such as CNT field-effect transistors (CNTFETs) and CNT diodes. CNTs can be used to form a *p–n* junction diode by chemical doping and polymer coating. These types of diodes can be used to form a computer chip. CNT diodes can potentially dissipate heat out of the   computer chips due to their unique thermal transmission properties (Jorio et al. 2008).

2.7.8 Interconnects

Carbon nanotubes (CNTs) have emerged as one of the most potential interconnect material solutions in current nanoscale regime. The higher current density of 1000 MA/sq-cm of an isolated CNT can eliminate the electromigration reliability concerns that plagues the current nanoscale copper interconnects. Therefore, CNT interconnects can potentially offer immense advantages over copper in terms of crosstalk, delay and power dissipation (Li et al. 2008b; Banerjee and Srivastava 2006a).

2.7.9 Transistors

CNTs can form conducting channels in transistor configurations as shown in Fig. 2.11. Two different device architectures have been developed for the transistor configuration. In both cases, CNTs connect the source and drain electrodes and show excellence behavior in the area of memory designing, amplifiers, sensors and detectors, etc. In one device architecture, the source and the drain are connected by a single nanotube. In other device architecture, a random array of nanotubes

Fig. 2.11 CNT-based transistor

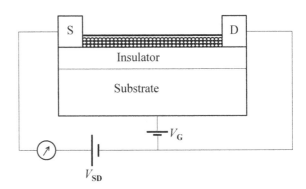

functions as a conducting channel (Gruner 2005; Banerjee and Srivastava 2006a; Raychowdhury and Roy 2006; Wind et al. 2002, 2003; Hasan et al. 2006; Appenzeller et al. 2005; Manney et al. 1992). The advantages of CNTFET over Si-MOSFET are as follows (Sahoo and Mishra 2009):

(i)   CNTFET demonstrates higher drive current compared to Si-MOSFET.
(ii) CNTFET shows approximately four times higher transconductance than Si-MOSFET.
(iii) The average carrier velocity of CNTFET is almost double the Si-MOSFET.

This chapter presented the unique atomic structures, properties and applications of carbon nanotubes. The electrical, mechanical and thermal properties of CNTs are primarily dependent on their diameter and chirality. In addition to this, the chapter summarized different production and purification methods for SWNTs and MWNTs.

Chapter 3
Modeling of Carbon Nanotube Interconnects

Abstract Modeling of a carbon nanotube interconnect is primarily dependent on its diameter, electron transport properties, chiralities, metallic and semiconducting properties. This chapter presents a technical review of analytical models for single-walled (SWNT), double-walled (DWNT), and multi-walled CNT (MWNT) structures. Depending on the geometry, the equivalent electrical models and the associated resistive, inductive, and capacitive parasitics for different SWNT, DWNT, and MWNT bundles are described.

Keywords Lüttinger liquid theory · Multi-conductor transmission line (MTL) · Multi-equivalent single conductor (MESC) · Equivalent *RLC* model · Quantum resistance · Kinetic inductance · Quantum capacitance

3.1 Introduction

After discovery of carbon nanotube in 1991 by scientist S. Iijima (Kase et al. 2003), extensive research on CNT interconnects has been carried out due to their unique electrical, mechanical, and thermal properties. For the first time, Burke (2002) modeled the CNT as nano transmission line using the Lüttinger liquid theory. Later on, several researchers reported the modeling of CNT interconnects based on their unique diameter-dependent properties, electron transport properties, chiralities, metallic, and semiconducting properties. This chapter presents different analytical models of single-, double-, and multi-walled CNT interconnects.

3.2 Analytical Models: A Technical Review

During recent past, several researchers have reported the modeling of single and bundled SWNT and MWNT based interconnects. This section presents a review of various analytical models proposed in past for different CNT structures.

© The Author(s) 2015
B.K. Kaushik and M.K. Majumder, *Carbon Nanotube Based VLSI Interconnects*, SpringerBriefs in Applied Sciences and Technology, DOI 10.1007/978-81-322-2047-3_3

3.2.1 Lüttinger Liquid Theory Based Model

Burke (2002) modeled nanotube as nanotransmission line with distributed kinetic and magnetic inductance as well as quantum and electrostatic capacitance using Lüttinger liquid collective modes. Later on, the Lüttinger liquid model was explained by Avouris et al. (2003) through extensive study of electronic structure and transport properties of CNTs. Based on the analysis, a bottom-up approach was reported by Li et al. (2003) to integrate MWNTs into multi-level interconnects in silicon-integrated circuits. Later, Miano and Villone (2005) extended this fluid theory model for frequency domain to describe electromagnetic response of three-dimensional (3D) structures formed by metallic CNTs and conductors within the framework of classical electrodynamics. Based on the extended liquid theory, Maffucci et al. (2009) presented a transmission line (TL) model to describe the propagation along SWNT interconnects for both isolated and bundled CNTs. Using this TL model, for small CNT radius, all the per unit length (*p.u.l.*) parameters were considered similar to models presented in Miano and Villone (2005); Salahuddin et al. (2005); and Wesstrom (1996). For large CNT radius, Salahuddin et al. (2005) and Wesstrom (1996) made the corrections for these *p.u.l.* parameters. Furthermore, a semiclassical one-dimensional (1D) electron fluid model was presented (Xu and Srivastava 2009) that takes into account electron–electron repulsive force. Based on the 1D electron fluid theory, the authors developed a transmission line model for metallic CNT interconnects by using classical electrodynamics.

Recently, Srivastava et al. (2010) reported different models of MWNT and bundled SWNT interconnects (shown in Figs. 3.1 and 3.2) based on the one-dimensional fluid theory. The equivalent electrical model of MWNT in Fig. 3.1a considers M number of shells where r_s, l_k, and c_q represent the *p.u.l.* scattering resistance, kinetic inductance, and quantum capacitance of each shell, respectively. The imperfect metal–nanotube contact resistance R_{mc} exhibits a typical value of 3.2 kΩ per shell depending on the fabrication process. The inter-shell coupling capacitance (c_s) and electrostatic capacitance (c_e) is primarily due to the potential difference between adjacent shells and the outer shell to ground plane, respectively. The simplified model of Fig. 3.1a is shown in Fig. 3.1b, c, wherein all the shells in MWNT are considered as parallel. Similarly, Fig. 3.2a presents the equivalent *RLC* model of SWNT bundle interconnects, where N_a and N_b are the number of SWNTs in ground and upper level, respectively. The inter-CNT bundle capacitance is represented by c_b. The total number of CNTs in bundle and the metallic to semiconducting ratio are represented as N and β, respectively. Considering all SWNTs in bundle as parallel, a simplified model is shown in Fig. 3.2b.

3.2.2 Electron Transport Theory Based Model

Several researchers reported different physical models of SWNT and MWNT interconnects based on the unique transport property of electrons. Ngo et al. (2004)

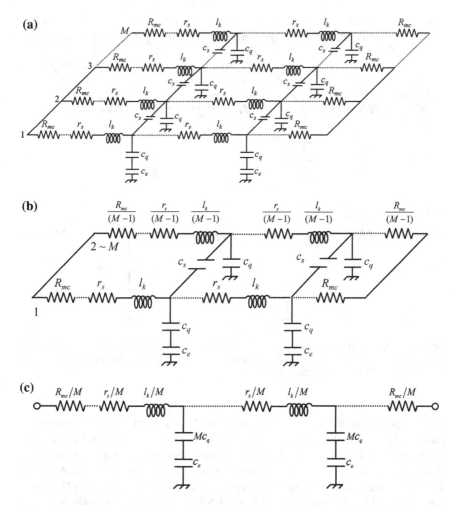

Fig. 3.1 **a** Equivalent electrical model of metallic MWNT interconnect. **b** Simplified equivalent circuit model of Fig. 3.1a for metallic MWNT interconnects, where 2 to M shells have been represented by single RLC equivalent line and the other outermost shell (*i.e.*, shell number 1) is represented by another RLC equivalent line. **c** Simplified equivalent circuit model of Fig. 3.1b for metallic MWNT interconnects

reported the mechanism of electron transport across metal CNT interface. The authors analyzed this mechanism for two different MWNT architectures, horizontal or side-contacted MWNTs and vertical or end-contacted MWNTs. Based on the physics of electron transport phenomenon, Naeemi and Meindl (2007a, b) presented an equivalent circuit model that captured changes in the number of conducting channels as a function of temperature. The model also takes into account various electron–phonon scattering mechanisms including emission and absorption of optical phonons and interaction with acoustic phonons. Recently, Li et al.

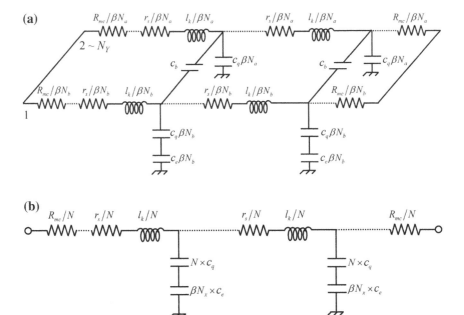

Fig. 3.2 **a** Equivalent circuit for a metallic SWNT bundle interconnects, where N_a and N_b represent number of SWNTs in ground and upper level, respectively. **b** Simplified equivalent circuit of Fig. 3.2a for SWNT bundle interconnects

(2009a) presented the current state of research in carbon based nanomaterials, particularly for one-dimensional (1D) form of CNTs and GNRs. A large number of attractive features such as electrical, thermal, mechanical properties and modeling techniques for CNTs were reviewed in this study (Li et al. 2009a). Again, on the basis of the electron transport properties, a novel transmission line (TL) model was presented for metallic and bundled SWNTs by Sarto et al. (2009), Sarto and Tamburrano (2010). Using closed form equations, the authors demonstrated per unit length charge deviation and current wave propagation equations in SWNT for the first time.

3.2.3 Models Based on Physical Parameters of CNTs

Similar to the electron transport property, various physical parameters such as length, diameter, conductivity, metallic, and semiconducting properties are also important for modeling of CNT interconnects. One of such compact physical models was reported for different number of conducting channels in MWNTs and SWNT bundles as well as their conductivities (Naeemi and Meindl 2006).

Depending on the physical properties, different resistances for bundled SWNTs and their impact for on-chip interconnect applications were modeled by Nieuwoudt and Massoud (2006a). The authors reported that the modeling of diameter dependent ohmic and contact resistances of individual SWNTs produce an error of 120 % or less compared to the previous case (Kim et al. 2005). Similar to the previous assumption, Nieuwoudt and Massoud (2006b) presented an accurate and scalable model for magnetic inductance of SWNT bundles considering the density and statistical distribution for both metallic and semiconducting nanotubes within the bundle. It has been observed that both the magnetic inductance and kinetic inductance are highly dependent on the bundle geometry. Moreover, a realistic *RLC* model of CNT interconnect is presented in Raychowdhury and Roy (2006). The authors reported that substantial numbers of metal layers are required for modeling of CNT interconnects to meet the performance of copper in same die area. Therefore, CNTs can provide reliable interconnect solution for future high-performance digital VLSI systems. Later, Agarwal et al. (2007) reported the existence of an energy barrier for thermally activated inter-CNT or inter-shell conduction within each CNT, in individual MWNTs and between neighboring CNTs in their random network using electrical transport measurement and density functional theory.

3.2.4 Diameter Dependent Modeling of CNT Interconnects

Haruehanroengra and Wang (2007) introduced a diameter dependent model to analyze the conductance of both SWNTs and MWNTs. The authors demonstrated that mixed CNT bundles (containing both SWNTs and MWNTs of different diameters) can provide two to five times conductance improvement over copper by selecting suitable parameters such as bundle width, tube density, and metallic tube ratio. Furthermore, based on the physical structures, distributed circuit models were presented (Naeemi and Meindl 2007b) for single and bundled SWNTs that are valid for all voltages and lengths. The work by Nieuwoudt and Massoud (2007) demonstrated the relative impact of magnetic and kinetic inductance on SWNT bundle interconnects. Based on the experimental results, it has been figured out that kinetic inductance will not be a significant factor in future bundled SWNT interconnect solutions. Moreover, Close and Wong (2008) developed a versatile method to make contact between array of individual MWNTs. To make suitable contact, comparisons were made between four different metals (Al, Au, Ti, and Pd) by incorporating the measurement of about 200 resistances. Following the versatile method, an accurate semi-analytical model was proposed (Kshirsagar et al. 2008) for intrinsic gate capacitance of CNT-array based back-gated field effect transistors (FET). The electrostatic screening effect in the FET is accounted for any given number of nanotubes, diameter, pitch, and gate dielectric thickness. Comprehensive modeling and designing techniques for CNT based interconnects were developed (Nieuwoudt and Massoud 2008) to examine the performance, reliability, and fabrication requirements for future nanotube based interconnect solutions. Simulation results at different technology nodes (14 nm, 22 nm, 32 nm) indicate that optimized nanotube bundles can provide significant reduction in delay than standard Cu wires and non-optimized MWNT and bundled SWNT.

3.2.5 Models Based on Process Induced Parameters

A new modeling technique has been presented (Fathi et al. 2009) for analyzing time-domain response of CNT interconnects based on transmission line modeling that takes into account the effects of process-induced contact resistance. The contact resistance of 14 different electrode metals with work function between 3.9 eV and 5.7 eV for CNT interconnects was investigated by Lim et al. (2009). It was observed that the contact resistance is mainly influenced by wettability and work function difference of electrode metal to CNT. Finally, a complete compact model of CNT interconnects was developed by Sinha et al. (2009). This model is scalable with process and design parameters and matches well with the numerical simulations as well as the measurement data.

3.2.6 Compact Physical Models of SWNT and MWNT Interconnects

Naeemi and Meindl (2005) observed that mono- or bi-layer SWNT interconnects above thick dielectric layers are promising candidates for short local interconnects because of their significantly smaller lateral capacitance compared to the conventional Cu wires. Using appropriate physical models, performance of SWNT and minimum-sized Cu interconnects were compared for various technology nodes of 18 nm, 22 nm, 38 nm, and 45 nm (Naeemi et al. 2005). It has been observed that at 22 nm technology node, nanotube bundles are 80 % faster than copper wires if electron mean free path in SWNTs is larger than the length of the nanotube (Naeemi and Meindl 2005). Koo et al. (2007) proposed a novel interconnect architecture of CNT bundle and compared it with Cu/low-k wires for future high-performance integrated circuits. The comparison demonstrated that for local wire, a CNT bundle exhibits smaller latency than Cu for a given geometry. Cho et al (2008) compared the performance of CNTs and optical interconnects with scaled Cu/low-k for future high-performance integrated circuits in terms of latency, power dissipation, bandwidth, and power density.

Li et al. (2008a) proposed a compact physical model of MWNT interconnect as shown in Fig. 3.3. In the equivalent circuit model, G_t represents the inter-shell tunneling conductance that is mainly due to the electron tunnel transport phenomenon between adjacent shells. Using the equivalent model, the performance of

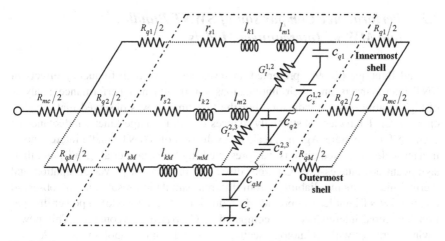

Fig. 3.3 Equivalent distributed circuit model of a MWNT with M number of shells

MWNTs is compared against traditional Cu- and SWNT- based interconnects at different technology nodes. They observed that at intermediate and global levels, MWNT interconnects can achieve smaller signal delay compared to Cu interconnects and these improvements became more pronounced with technology scaling and longer interconnects.

The problem of scaling interconnects to nanometric dimensions in future VLSI applications was first addressed by Maffucci et al. (2008a, b). Based on the electron liquid theory (Burke 2002), a transmission line (TL) model was proposed to describe the propagation along SWNT. The TL model was further extended to multi-conductor structures and applied to describe the behavior of nanotube bundles. Furthermore, some physical models were presented (Naeemi and Meindl 2008) to benchmark CNTs (SWNTs/MWNTs) against Cu wires for both signal and power distribution at realistic on-chip temperature (100 °C). The simulated results in Naeemi and Meindl (2008) indicate that Cu wires, SWNTs, and MWNTs have distinct advantages and disadvantages, and therefore, a hybrid system made out of Cu/SWNT/MWNT interconnects would be optimal for multi-level interconnect network. Based on physical models, some drawbacks of Cu/low-k technology and a conventional repeater inserted signaling scheme for global interconnects were discussed by Koo et al. (2009). To overcome these problems, some alternatives were proposed on the basis of CNTs, optics, and a new circuit scheme named as "Capacitively Driven Low-Swing Interconnect (CDLSI)." The performance characteristics for densely packed SWNT bundles were investigated for semi-global interconnects (Giustiniani et al., 2010). To analyze the delay performance, the virtual effects for insertion of variable number of repeaters along the interconnect and their influence on contact resistance between CNTs and external circuitry were also studied. It was observed that the reduction of delay for CNT based interconnects was not at par with Cu based interconnects for semi-global level.

3.2.7 Performance Comparison of SWNT Bundles and MWNT Interconnect Models

Li and Banerjee (2009b) presented an investigation of high-frequency effects in CNT interconnects and their implications for design and performance analysis of high quality on-chip inductors. For the first time, it was observed that the skin effect in CNT bundles substantially reduces due to the large kinetic inductance of each CNT in a bundle. Again, a realistic evaluation of SWNT bundle interconnects in nanoscale-integrated circuits was presented by Srivastava et al. (2009) and their significant advantages over Cu were discussed in terms of power dissipation and thermal management/reliability. From different simulation results, it was observed that the SWNT bundles can provide 4x reduction (at 22 nm node) in power dissipation for global interconnects as compared to Cu based interconnects. This power saving increases with technology scaling to 8x at 14 nm technology node. A compact physical model was developed (Belluci and Onorato 2010) for determining the ultimate potential performance of MWNTs, with various diameters and lengths, and compared them with Cu wires and SWNT bundles. A new approach was developed to obtain the electron spectrum of metallic MWNT with an arbitrary number of coaxial layers by inter-tube coupling. However, a comparative analysis was reported (Pasricha et al. 2010) to demonstrate the performance and energy impact for CNT based global interconnects over Cu using multi-core chip multiprocessor (CMP) applications. Experimental results indicate that SWNTs are not suitable for global interconnect buses due to their higher ohmic resistance and delay, whereas global MWNT buses can provide performance speedups of up to 1.98x for CMP applications. The performance gain can be further improved by improving CNT mean free path. Recently, Jamal and Naeemi (2011) demonstrated that single isolated SWNTs are promising candidate for interconnects in subthreshold circuits because of their lower capacitance. Due to large resistance of individual SWNTs, bundles of densely packed SWNTs were proposed to be used as interconnects.

3.2.8 CNT Interconnect Models for FPGA Applications

One of the most interesting comparisons based on the practical application of CNT and Cu was reported by Eachempati et al. (2009) . The authors investigated the performance and reliability of routing architectures in field effect programmable gate arrays (FPGA) that utilized SWNT bundles as wires in the FPGA interconnect fabric for future process technologies. Investigation results demonstrated that FPGAs utilizing SWNT bundle interconnect can achieve upto 54 % improvement in area-delay product over the best performing architecture with standard Cu in 22 nm process technology. Based on the theoretical analysis, advantages of carbon nanomaterials like CNTs and GNRs were reported in all domains that include local and global interconnects, vias, through silicon vias (TSVs), and off-chip interconnections (Li et al. 2010). It has been reported that both the CNTs and GNRs can reduce interconnect delay and power consumption upto 60 % and 50 %, respectively, for global interconnects.

3.2.9 CNT Interconnect Models for Crosstalk Analysis

Propagation delay under crosstalk influence is an important design concern in modern VLSI interconnects. For the first time, Rossi et al. (2007) reported the dynamic crosstalk effect in CNT bus architecture that was implemented by parallel SWNTs or a single MWNT as shown in Fig. 3.4a, b, respectively. The coupling capacitance (c_{cm}) primarily depends on the distance between adjacent lines and the diameters of CNTs. The authors proposed a novel bus architecture with low crosstalk feature, wherein the DWNTs are arranged parallelly.

Another study on crosstalk effect was reported by Chen et al. (2009). In the proposed model, the hybrid effects of several geometrical and physical parameters of SWNT interconnect arrays were numerically investigated depending on their temperature distribution, breakdown voltage, power handling capabilities, and transient thermal response. The crosstalk effects in SWNT and DWNT bundled interconnect architecture were investigated by Pu et al. (2009) based on the information provided in ITRS 2006. Figure 3.5a, b presents the equivalent electrical models of SWNT and DWNT bundles, wherein n_{CNT} number

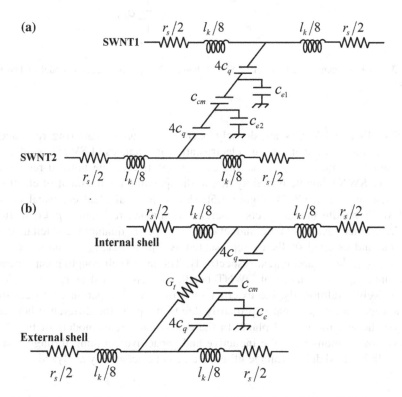

Fig. 3.4 **a** Equivalent *RLC* circuit model for parallel SWNTs. **b** Equivalent *RLC* circuit model for a single MWNT

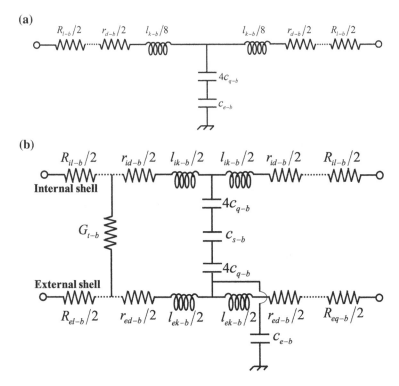

Fig. 3.5 a Equivalent circuit model of SWNT bundle. **b** Equivalent circuit model of DWNT bundle

of SWNTs and DWNTs are densely packed. The *p.u.l.* scattering resistance, kinetic inductance, quantum and electrostatic capacitances of SWNT bundle are represented as r_{d-b}, l_{k-b}, c_{q-b}, and c_{e-b}, respectively. The lumped resistance (R_{l-b}) of SWNT bundle is mainly due to the quantum confinement of electrons in a nanowire (i.e., CNT). Figure 3.5b shows the equivalent electrical model of DWNT bundle interconnects, wherein the lumped resistance, *p.u.l.* scattering resistance, *p.u.l.* kinetic inductance, and *p.u.l.* quantum capacitance of internal and external shells are represented as R_{il-b}, r_{id-b}, l_{ik-b}, and c_{iq-b}, and R_{el-b}, r_{ed-b}, l_{ek-b}, and c_{eq-b}, respectively. The inter-shell coupling capacitance and tunneling conductance of DWNT bundle are represented as c_{s-b} and G_{t-b}, respectively. Additionally, the external shell of DWNT experiences an electrostatic capacitance c_{e-b} that is primarily due to the potential difference between the bundle and the ground plane. In the equivalent circuit models of Fig. 3.5, the authors demonstrated the inductive and capacitive coupling through which crosstalk induced delay, voltage glitch, etc., can be accurately analyzed.

3.2.10 Modeling of Mixed CNT Bundle Interconnects

Modeling and simulation of SWNT and MWNT interconnects have already become popular for analyzing the propagation delay, power dissipation and crosstalk performance (Li et al. 2008; Pu et al. 2009; Amore et al. 2010). Several researchers based on CNT fabrication outcome reported that a realistic nanotube bundle contains not only SWNTs but also the MWNTs, that too of different diameters. Moreover, in current research scenario, it is preferred to fabricate and model the bundles with CNTs of different diameters, so as to achieve a densely packed mixed CNT bundle (MCB). Recently, Sathyakam and Mallick (2011) presented a hierarchical modeling approach for mixed CNT bundle (mixture of SWNT and MWNT) interconnects. They reported that mixed CNT bundle is superior to MWNT or SWNT bundle interconnects in terms of delay. Subash et al. (2013) reported a significant reduction in capacitive crosstalk using an MCB wherein MWNTs are placed at periphery to the centrally located SWNTs.

3.3 Geometry and Equivalent *RLC* Model of CNT Interconnect

This section presents an equivalent electrical model of single and bundled SWNT, DWNT, and MWNT interconnect. Depending on the geometry, distributed *RLC* parameters and their equivalent circuit models are as follows:

3.3.1 SWNT Interconnect

Consider a SWNT above the ground plane as shown in Fig. 3.6a. The SWNT of diameter D is placed above the ground plane at a distance h_t. Therefore, the distance between the center of SWNT and the ground plane is $H = (h_t + D/2)$. Depending on the geometry, an equivalent electrical circuit model of SWNT interconnect is shown in Fig. 3.6b.

The interconnect parasitics of the equivalent model are as follows:

A. **Resistance**

In the equivalent model of Fig. 3.6b, r_s represents the scattering resistance of SWNT that appears due to the static impurity scattering, defects, line edge roughness scattering (LER), etc. The scattering resistance primarily depends on the nanotube length (l) and power supply voltage (V_{dd}) and can be expressed as (Rossi et al. 2007)

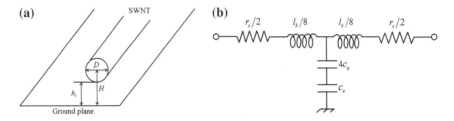

Fig. 3.6 **a** Geometry and, **b** equivalent *RLC* model of SWNT interconnect

$$r_s = \frac{h}{4e^2} \quad \text{if } l < \lambda_{\text{high}}$$

$$= \frac{h}{4e^2} \left[\frac{1}{0.64(l - \lambda_{\text{high}}) + \lambda_{\text{high}}} \right] \quad \text{if } \lambda_{\text{high}} < l < \lambda_{\text{low}} \qquad (3.1)$$

$$= \frac{h}{4e^2} \left[\frac{1}{0.64(\lambda_{\text{low}} - \lambda_{\text{high}}) + \lambda_{\text{high}}} \right] \quad \text{if } l > \lambda_{\text{low}}$$

where h, e, λ_{high} and λ_{low} are Planck's constant, electron charge, mean free path of backscattering for high and low biases, respectively. For nanotube length smaller than the mean free path of electrons, the resistance is length independent and exhibits the ballistic phenomenon. However, for longer interconnects, the electron transport is not ballistic since the increase in resistance is caused by scattering mechanisms.

B. Inductance

SWNT primarily demonstrates two different types of inductances: (1) kinetic and (2) magnetic inductances. The kinetic inductance (l_k) is mainly due to the kinetic energy of electrons along the nanotube length and can be expressed as

$$l_k = \frac{h}{2e^2 v_F} \qquad (3.2)$$

where v_F is the Fermi velocity of graphene and CNT $\approx 8 \times 10^5$ m/s (Burke 2002). From Eq. (3.2), l_k can be approximated as 16 nH/μm. Due to four propagating channels associated with SWNT, the effective kinetic inductance is $l_k/4$. Apart from this, the magnetic inductance of SWNT is due to the stored energy of carriers in magnetic field. However, due to the smaller value (in pH range), the equivalent model of Fig. 3.6b neglects the magnetic inductance.

C. Capacitance

The equivalent *RLC* model of Fig. 3.6b consists of two types of capacitances:

(1) *Electrostatic capacitance*: It represents the electrostatic coupling between two neighboring CNTs or between a CNT and ground plane.

(2) *Quantum capacitance*: It originates from the finite density of states at Fermi energy, i.e., a finite amount of energy required to add an extra electron above the Fermi level.

The electrostatic capacitance (c_e) between nanotube and the ground plane can be expressed as (Rossi et al. 2007)

$$c_e = \frac{2\pi\varepsilon_0\varepsilon_r}{\cosh^{-1}(2H/D)} \tag{3.3}$$

where $\varepsilon_r \approx 2.2$ is the relative permittivity of the medium between SWNT and the ground plane. For typical values of H/D, c_e can be approximated as 50 aF/μm (Rossi et al. 2007). The quantum capacitance (c_q) in *p.u.l.* can be expressed as

$$c_q = \frac{2e^2}{hv_F} \tag{3.4}$$

The typical value of c_q is 100 aF/μm (Burke 2002). Due to four propagating channels in parallel, the effective quantum capacitance in the equivalent circuit is $4c_q$.

3.3.2 DWNT Interconnect

DWNT is the simplest geometry of an MWNT, wherein only two shells are present. Due to the Vander Waal's force between neighboring carbon atoms, the minimum distance (δ) between two shells is 0.34 nm. The diameters of internal and external shells are represented as D_{in} and D_{out}, respectively, as shown in Fig. 3.7a. The distance between the center of DWNT and the ground plane is $H = (h_t + D_{out}/2)$.

Depending on the geometry, an equivalent *RLC* model of DWNT interconnect is presented in Fig. 3.7b. The dependability of these interconnect parasitics and brief explanation about them is provided as follows:

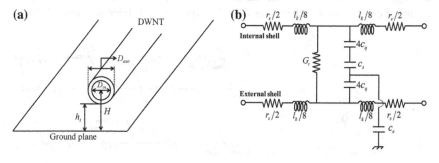

Fig. 3.7 a Geometry and, **b** equivalent *RLC* model of DWNT interconnect

A. **Resistance**

The scattering resistance (r_s) of each shell in DWNT is similar to that of a single SWNT and can be expressed as (3.1). Additionally, a tunneling conductance $G_t \approx (10 \text{ k}\Omega)^{-1}/\mu\text{m}$ exists between the adjacent shells of DWNT that is due to the inter-shell electron tunnel transport phenomenon (Bourlon et al. 2004).

B. **Inductance**

Kinetic inductance (l_k) of each shell in DWNT is similar to that of single SWNT and can be expressed as (3.2). However, due to the smaller value, magnetic inductance has not been considered in the *RLC* circuit of Fig. 3.7b.

C. **Capacitance**

Quantum capacitance (c_q) of each shell in DWNT is also similar to that of single SWNT interconnects and can be expressed as (3.4). However, the electrostatic capacitance (c_e) appears only between the external shell and the ground plane since the outer shell of DWNT shields the internal ones. The c_e between the outer shell of DWNT and the ground plane can be expressed as (Rossi et al. 2007)

$$c_e = \frac{2\pi \varepsilon_0 \varepsilon_r}{\cosh^{-1}\left(2H/D_{out}\right)} \tag{3.5}$$

Due to metallic behavior of DWNT interconnect, the inter-shell coupling capacitance (c_s) can be expressed as

$$c_s = \frac{2\pi \varepsilon_0}{\ln\left(D_{out}/D_{in}\right)} \tag{3.6}$$

3.3.3 MWNT Interconnect

MWNT has multiple number of shells. Consider the geometry of MWNT above ground plane as shown in Fig. 3.8a. The inner and outer shell diameters of MWNT are referred as D_1 and D_n, respectively. l and δ denote the nanotube length and inter-shell spacing, respectively. H represents the distance between the center of MWNT and the ground plane. The inter-shell spacing δ can be expressed as (Majumder et al. 2012a)

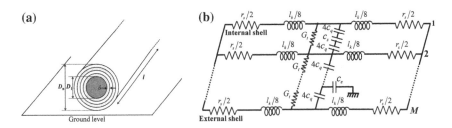

Fig. 3.8 **a** Geometry and, **b** equivalent *RLC* model of single MWNT interconnect

$$\delta = \frac{D_n - D_{n-1}}{2} \approx 0.34 \text{ nm} \tag{3.7}$$

Depending on the geometry, an equivalent *RLC* model of MWNT interconnect is presented in Fig. 3.8b, wherein each shell of MWNT is modeled as nanotransmission line. The interconnect parasitics of MWNT are as follows:

A. Resistance

The scattering resistance (r_s) of each shell in MWNT is similar to that of SWNT/DWNT and can be expressed as (3.1). Additionally, an inter-shell tunneling conductance $G_t \approx (10 \text{ k}\Omega)^{-1}/\mu\text{m}$ is primarily due to the electron tunnel transport phenomenon between neighboring shells (Li et al. 2008b).

B. Inductance

The kinetic inductance (l_k) of each shell in MWNT is similar to that of SWNT/DWNT and can be expressed as (3.2).

C. Capacitance

The quantum capacitance (c_q) of each shell in MWNT can be computed similar to the SWNT/DWNT and expressed as (3.4). The electrostatics capacitance (c_e) appears between the external shell of MWNT and the ground plane, as external shell shields the internal ones and can be expressed as (3.5). Electrostatic coupling capacitance (c_s) between adjacent shells of MWNT can be expressed as (3.6).

3.3.4 SWNT Bundle Interconnect

The geometry of SWNT bundle with height h and width w is shown in Fig. 3.9a. The bundle contains number of SWNTs of diameter D. The center-to-center distance between neighboring SWNTs in bundle is $S_{C-C} = (D + \delta)$, where $\delta = 0.34$ nm represents the Vander Waal's distance between neighboring carbon atoms. The total number of SWNTs can be obtained as (Pu et al. 2009; Majumder et al. 2012a)

$$n_{SWNT} = \begin{cases} n_W n_H - (n_H/2) & \text{if } n_H \text{ is even} \\ n_W n_H - (n_H - 1/2) & \text{if } n_H \text{ is odd} \end{cases} \tag{3.8}$$

where

$$n_W = \left\lfloor \frac{w - D}{S_{C-C}} \right\rfloor + 1; \; n_H = \left\lfloor \frac{h - D}{S_{C-C}} \right\rfloor + 1 \tag{3.9}$$

n_H and n_W are number of rows and columns in the bundle, respectively, and n_{SWNT} is the total number of SWNTs in bundle. Depending on the bundle geometry, an equivalent *RLC* model of SWNT bundle is shown in Fig. 3.9b. This section presents a brief description of equivalent bundle resistance, inductance, and capacitance.

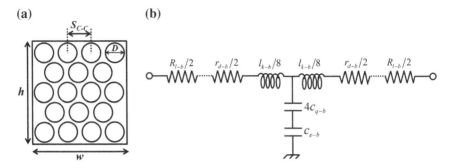

Fig. 3.9 **a** Geometry and, **b** equivalent *RLC* model of SWNT bundle interconnect

A. Bundle Resistance

Equivalent resistance (R_{CNT}) of SWNT bundle primarily consists of (1) quantum or intrinsic resistance (R_q) that is due to the quantum confinement of electrons in a nano-wire, (2) imperfect metal-nanotube contact resistance (R_c) that exhibits a value of few to several hundreds of kilo-ohms depending on the fabrication process and (3) scattering resistance (r_s) that appears due to the static impurity scattering, defects, line edge roughness scattering (LER) and acoustic phonon scattering (Pu et al. 2009; Majumder et al. 2012a). The R_{CNT} can be expressed as

$$R_{CNT} = R_c + R_q + r_s \cdot l_{CNT}$$

$$= R_c + R_q\left(1 + \frac{l_{CNT}}{\lambda_{mfp}}\right) \qquad (3.10)$$

where λ_{mfp} and l_{CNT} represent the mean free path and interconnect length, respectively. For the bundle interconnects, since all the SWNTs are connected in parallel, the bundle resistance can be expressed as (Pu et al. 2009)

$$R_b = \frac{R_{CNT}}{n_{SWNT}} = \frac{(R_c + R_q)}{n_{SWNT}} + \frac{R_q l_{CNT}}{n_{SWNT} \cdot \lambda_{mfp}} \qquad (3.11)$$

$$= R_{l-b} + r_{d-b} l_{CNT}$$

where $R_{l-b} = \frac{(R_c + R_q)}{n_{SWNT}}$ and $r_{d-b} = \frac{R_q}{n_{SWNT} \cdot \lambda_{mfp}}$ represents the lumped and distributed resistances of the SWNT bundle, respectively.

B. Bundle Inductance

The kinetic inductance mainly depends on the number of SWNTs in the bundle and can be expressed as

$$l_{k-b} = l_k / 4 n_{SWNT} \qquad (3.12)$$

where l_k is the kinetic inductance of individual SWNT and can be expressed as (3.2). Due to the dominating value of kinetic inductance (in nH range), the magnetic or mutual inductance of a SWNT bundle is neglected in the equivalent model shown in Fig. 3.9b.

C. Bundle Capacitance

The equivalent capacitance of SWNT bundle primarily consists of (1) quantum capacitance (c_{q-b}) that represents the density of electronic states and (2) electrostatic capacitance (c_{e-b}) that is mainly due to the potential difference between the bundle and the ground plane. The quantum capacitance of a bundle mainly depends on the total number of SWNTs and can be expressed as (Pu et al. 2009)

$$c_{q-b} = c_q.n_{SWNT} \tag{3.13}$$

Due to the four conducting channels of SWNT, the equivalent quantum capacitance of the bundle is equal to $4c_{q-b}$. The electrostatic capacitance of SWNT bundle can be obtained as (Pu et al. 2009)

$$c_{e-b} = \frac{2\pi\varepsilon_0\varepsilon_r}{\ln(H/D)} \times n_x \tag{3.14}$$

where n_x represents the number of SWNTs facing the ground plane.

3.3.5 DWNT Bundle Interconnect

The geometry of DWNT bundle is shown in Fig. 3.10a, wherein the inner and outer diameters are referred as D_{in} and D_{out}, respectively. The distance between two DWNTs in bundle is $S_{C-C} = (D_{out} + \delta)$. Therefore, replacing the diameter term D by external shell diameter D_{out} in (3.9), the total number of DWNTs (n_{DWNT}) in a bundle can be easily obtained. Based on the geometry, an equivalent electrical model of DWNT bundle is presented in Fig. 3.10b.

The interconnect parasitics of DWNT bundle are described below:

A. Bundle Resistance

The diameter of internal shell is different from its external shell; thus, their corresponding *p.u.l.* resistances are different and denoted as R_{int} and R_{ext}, respectively. The series internal and external resistances of DWNT bundle can be expressed as

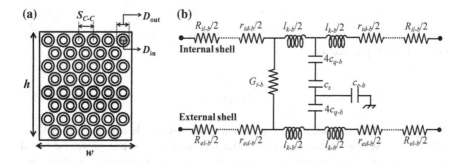

Fig. 3.10 **a** Geometry and, **b** equivalent *RLC* model of bundled DWNT interconnect

$$R_{i-b} = \frac{R_{\text{int}}}{n_{DWNT}} = \frac{(R_c + R_{qi})}{n_{DWNT}} + \frac{R_{qi}l_{CNT}}{n_{DWNT} \cdot \lambda_{mfp}} \tag{3.15}$$
$$= R_{il-b} + r_{id-b}l_{CNT}$$

$$R_{e-b} = \frac{R_{\text{ext}}}{n_{DWNT}} = \frac{(R_c + R_{qe})}{n_{DWNT}} + \frac{R_{qe}l_{CNT}}{n_{DWNT} \cdot \lambda_{mfp}} \tag{3.16}$$
$$= R_{el-b} + r_{ed-b}l_{CNT}$$

where $R_{il-b} = \frac{(R_c + R_{qi})}{n_{DWNT}}$ and $R_{el-b} = \frac{(R_c + R_{qe})}{n_{DWNT}}$ represents the lumped resistance of internal and external shells, respectively. Similarly, the distributed resistance of internal and external shells are represented as $r_{id-b} = \frac{R_{qi}}{n_{DWNT} \cdot \lambda_{mfp}}$ and $r_{ed-b} = \frac{R_{qe}}{n_{DWNT} \cdot \lambda_{mfp}}$ respectively. Additionally, a tunneling conductance $G_t \approx (10k\Omega)^{-1}/\mu m$ exists between adjacent shells in DWNTs. This conductance is strongly affected by electron tunneling phenomenon between neighboring shells (Dadgour et al. 2008). The bundle conductance (G_{t-b}) in terms of G_t and number of DWNTs can be expressed as

$$G_{t-b} = G_t \cdot n_{DWNT} \tag{3.17}$$

B. Bundle Inductance

The kinetic inductance (l_{k-b}) of the bundled DWNT is similar to the bundled SWNT and can be expressed as

$$l_{k-b} = \frac{l_k}{4n_{DWNT}} \tag{3.18}$$

The magnetic inductance is neglected in the *RLC* model of Fig. 3.10b due to the dominating effect of kinetic inductance.

C. Bundle Capacitance

The quantum capacitance (c_{q-b}) and electrostatic capacitance (c_{e-b}) of the bundled DWNT are similar to the bundled SWNT and can be expressed as

$$c_{q-b} = c_q \cdot n_{DWNT} \tag{3.19}$$

Due to metallic behavior of DWNT interconnects, the inter-shell coupling capacitance can be expressed as

$$c_s = \frac{2\pi \varepsilon_0}{\ln (D_2/D_1)} \tag{3.20}$$

This chapter discussed electrical equivalent models of single and bundled SWNT, DWNT and MWNT interconnects. A brief technical review was presented based on their unique diameter-dependent properties, electron transport properties, chiralities, metallic, and semiconducting properties.

Chapter 4
Crosstalk and Delay Analysis

Abstract This chapter analyzes and compares the power, delay, and crosstalk performance of different CNT based interconnects in form of single and bundled SWNT, DWNT, and MWNT. The equivalent electrical models of CNT interconnects represent the interconnect line of a driver-interconnect-load (DIL) setup. The motivation behind using CMOS driver to drive the interconnect line is briefly discussed. The DIL setup is primarily used to analyze the propagation delay, whereas a capacitively coupled three-line bus architecture is used to analyze the crosstalk induced delay. The propagation delay under the influence of crosstalk is analyzed for different interconnect lengths, transition time, and spacing between coupled lines (aggressor and victim).

Keywords Propagation delay · Crosstalk induced delay · Driver-interconnect-load (DIL)

4.1 Introduction

This chapter provides a detailed analysis of propagation delay under the crosstalk influence for single and bundled SWNT, DWNT, and MWNT interconnects using capacitively coupled three-line bus architecture. Crosstalk noise in coupled lines can be broadly classified into two categories: (1) functional and (2) dynamic crosstalk (Kaushik et al. 2010). Under functional crosstalk category, victim line experiences a voltage spike when the aggressor line switches (Rabaey 2002). On the other hand, dynamic crosstalk is observed when aggressor and victim line switches simultaneously. A change in signal propagation delay is experienced under dynamic crosstalk when adjacent line (aggressor) switches either in same direction (in phase) or in opposite direction (out phase).

B.K. Kaushik and M.K. Majumder, *Carbon Nanotube Based VLSI Interconnects*, SpringerBriefs in Applied Sciences and Technology, DOI 10.1007/978-81-322-2047-3_4

4.2 Simulation Setup

This section briefs about the estimation of crosstalk induced delay using capacitively coupled three-line bus architecture. A novel concept of CMOS driver is introduced for accurate estimation of propagation delay under crosstalk influence.

4.2.1 Motivation Behind Using CMOS Driver

Accurate estimation of crosstalk waveform shape and peak noise of driver-interconnect-load (DIL) system is an important design concern since a long time (Kaushik and Sarkar 2008). Previously, several researchers modeled crosstalk noise in distributed capacitive and inductive coupled *RLC* interconnects using resistive driver. The use of resistive driver instead of CMOS driver generates a discrepancy in results. It can be understood by noting the fact that a transistor in CMOS gate operates partially in linear region and partially in saturation region during switching. But, a transistor can be accurately approximated by resistor only in the linear region. In the saturation region, the transistor is more accurately modeled as current source with parallel high resistance. Like other authors, Agarwal et al. (2006) incorrectly assumed the driver impedance to be a linear resistance. Similar incorrect analysis has also been carried by Rossi et al. (2007) and Pu et al. (2009) for bundled SWNT and MWNT interconnects.

For justifying the motivation, HSPICE simulations have been performed for two sets of *RLC* interconnects using a resistive and CMOS driver as shown in Fig. 4.1a, b, respectively. Simulations are run for different local and global interconnect lengths ranging from 100 nm to 1 μm and 100 μm to 1 mm, respectively (Majumder et al. 2012b). Table 4.1 demonstrates the propagation delay for different interconnect lengths using the resistive and the CMOS drivers. It has been observed that delay under crosstalk influence is increased by 80 % for the case of resistive driver compared to the CMOS driver. Therefore, a resistive driver can lead to severe errors while estimating crosstalk induced delay.

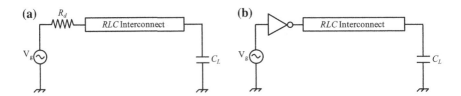

Fig. 4.1 A driver-interconnect-load (*DIL*) system with **a** resistive driver and **b** CMOS driver

Table 4.1 Propagation delay using resistive and CMOS driver for different interconnect lengths

Type of interconnects	Interconnect length (μm)	Propagation delay (in ps) of DIL system using	
		Resistive driver	CMOS driver
Local	0.1	0.76	0.11
	0.5	1.63	0.25
	1	2.75	0.43
Global	100	1149.50	1019.4
	500	25,567	24,955
	1,000	100,810	99,583

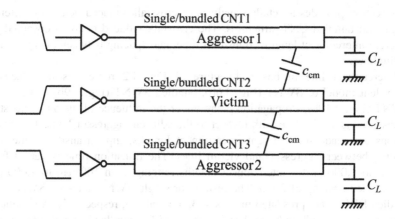

Fig. 4.2 Bus architecture made of three parallel CNTs (single/bundle SWNT and bundled DWNT)

4.2.2 Simulation Setup Using Capacitively Coupled Three-Line Bus Architecture

Propagation delay under the crosstalk influence has been analyzed for different interconnect models using a capacitively coupled three-line bus architecture (Rossi et al. 2007) as shown in Fig. 4.2. Out of these three lines, two are referred as aggressors and the middle one as victim. The interconnect line in bus architecture is represented by equivalent RLC models of different CNT structures as discussed in Chap. 3. Each line in the bus architecture is driven by CMOS driver. The coupling capacitance (c_{cm}) demonstrates the effect of crosstalk induced delay that primarily depends on the spacing (S_p) between aggressor and victim lines and can be expressed as (Majumder et al. 2012a)

$$c_{cm} = \frac{\pi \varepsilon_0 \varepsilon_r}{\cosh^{-1}\left(S_p / d_{avg}\right)} \tag{4.1}$$

where d_{avg} is the average diameter of CNTs facing each other. The bus architecture has the following values of load capacitance (C_L) and power supply voltage (V_{dd}): $C_L = 10$ aF and $V_{dd} = 1$ V. HSPICE simulations are performed for simultaneous signals at aggressor line when victim line is switched in opposite direction (out phase). Crosstalk induced delay is analyzed for different global interconnect lengths ranging from 100 µm to 1,000 µm.

4.3 Crosstalk Induced Delay of Bundled SWNT and DWNT Interconnects

This section provides a detailed analysis of crosstalk induced delay for different single and bundled CNT structures. The crosstalk induced delay is observed for different interconnect lengths, transition time, and spacing between aggressor and victim lines.

Interconnect line in bus architecture of Fig. 4.2 represents the electrical equivalent model of SWNT (Fig. 3.6b), bundled SWNT (Fig. 3.9b) and bundled DWNT (Fig. 3.10b). Propagation delay under the influence of dynamic crosstalk is observed at victim line with respect to the adjacent aggressor lines. The observations are made for different interconnect lengths, input transition time, and spacing between aggressor and victim lines. The transition time is varied from 100 ps to 1,000 ps for interconnect lengths ranging from 100 µm to 1,000 µm with a fixed spacing of 2 nm. The results for single SWNT, bundled SWNT, and bundled DWNT are presented in Figs. 4.3, 4.4 and 4.5, respectively. A significant reduction in crosstalk induced delay is observed for bundled DWNT compared to single or bundled SWNT interconnects. Figure 4.6 demonstrates a similar observation for different interconnect lengths at 1 ns transition time and 2 nm spacing. Finally, the crosstalk induced delay for different spacing ranging from 2 nm to 5 nm at fixed transition time of 1 ns and interconnects length of 100 µm is shown in Fig. 4.7.

Fig. 4.3 Crosstalk induced delay of SWNT for different transition time at 2 nm spacing

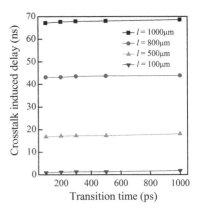

Fig. 4.4 Crosstalk induced delay of SWNT bundle for different transition time at 2 nm spacing

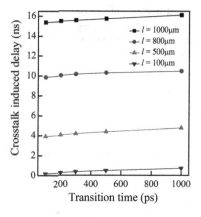

Fig. 4.5 Crosstalk induced delay of DWNT bundle for different transition time at 2 nm spacing

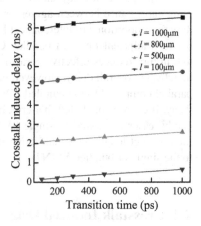

Fig. 4.6 Crosstalk induced delay for different interconnect lengths at 1 ns transition time and 2 nm spacing

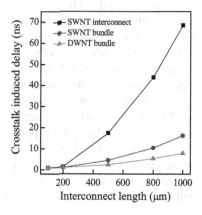

Fig. 4.7 Crosstalk induced
delay for different spacing
at 1 ns transition time and
100 μm length

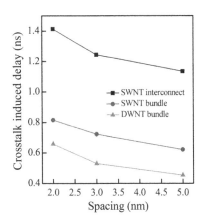

The propagation delay under crosstalk influence primarily depends on the coupling capacitance between aggressor and victim lines. The coupling capacitance is
directly proportional to interconnect length and inversely proportional to the spacing
between adjacent lines. In bundled CNT, all SWNTs and DWNTs are in parallel
that in turn reduces effective parasitic components (i.e., resistance, inductance) and
thereby, the overall delay. However, bundled DWNTs have more number of shells in
parallel compared to the bundled SWNTs that further reduces the crosstalk induced
delay. It can be concluded that crosstalk induced delay for bundled CNTs is appre-
ciably lesser compared to single SWNT. For a fixed transition time, spacing, and
interconnect length, crosstalk induced delay of bundled DWNT is lesser compared
to the single or bundled SWNTs.

4.4 Crosstalk Induced Delay of Bundled SWNT
and Single MWNT Interconnects

Crosstalk induced delay is analyzed in this section for different global interconnect lengths at fixed transition time and spacing of 1 ns and 2 nm, respectively.
Figures 4.8, 4.9, 4.10, 4.11 and 4.12 show the equivalent number of SWNTs in bundle for fixed number of shells in MWNTs, whereas Fig. 4.13 summarizes all the
results from Figs. 4.8, 4.9, 4.10, 4.11 and 4.12. It is observed that crosstalk induced
delay increases for longer interconnects whereas the delay reduces for more number
of shells in MWNT and number of SWNTs in a bundle (Majumder et al. 2012b).

Table 4.2 summarizes the number of SWNTs required in a bundle that achieves
same crosstalk induced delay performance with given number of shells in MWNT.
Similarly, Table 4.3 shows the equivalency between SWNTs and MWNTs with
corresponding areas for achieving similar crosstalk induced delay performance.
Based on the results provided in Tables 4.2 and 4.3, percentage reduction in crosstalk induced delay and area are summarized in Tables 4.4 and 4.5, respectively.

Fig. 4.8 Equivalent number
of SWNTs in bundle for fixed
number of shells in MWNTs
for similar performance of
crosstalk induced delay at
100 μm interconnects length

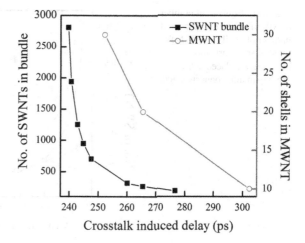

Fig. 4.9 Equivalent number
of SWNTs in bundle for fixed
number of shells in MWNTs
for similar performance of
crosstalk induced delay
at 200 μm interconnects length

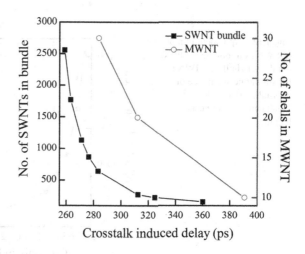

Fig. 4.10 Equivalent number
of SWNTs in bundle for fixed
number of shells in MWNTs
for similar performance of
crosstalk induced delay at
500 μm interconnects length

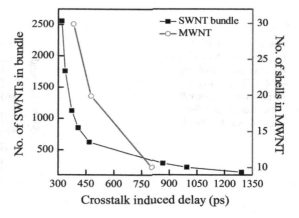

Fig. 4.11 Equivalent number
of SWNTs in bundle for fixed
number of shells in MWNTs
for similar performance
of crosstalk induced delay
at 800 μm interconnects length

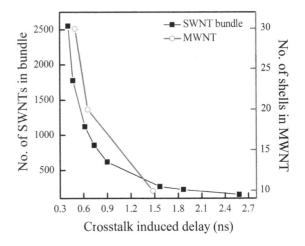

Fig. 4.12 Equivalent number
of SWNTs in bundle for fixed
number of shells in MWNTs
for similar performance of
crosstalk induced delay at
1,000 μm interconnects
length

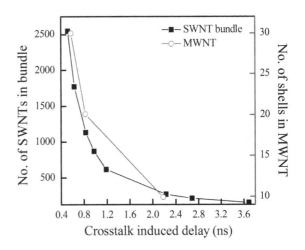

Fig. 4.13 Equivalent
number of SWNTs in bundle
for given number of shells
in MWNTs for similar
performance of crosstalk
induced delay at different
interconnects lengths (*l*)

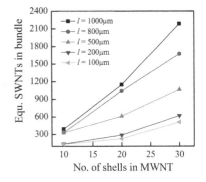

Table 4.2 Equivalent number of SWNTs in bundle with respect to fixed number of shells in MWNT for similar crosstalk induced delay

Interconnect lengths (μm)	Number of shells in MWNT = 10		Number of shells in MWNT = 20		Number of shells in MWNT = 30	
	Eqv. no. SWNT	Delay (ns)	Eqv. no. SWNT	Delay (ns)	Eqv. no. SWNT	Delay (ns)
100	127	0.302	232	0.265	505	0.252
200	138	0.390	274	0.311	610	0.283
500	326	0.782	610	0.468	1,061	0.377
800	337	1.476	1050	0.656	1,670	0.493
1,000	384	2.176	1145	0.827	2,185	0.572

Table 4.3 Demonstration of area for MWNTs with respect to equivalent SWNTs in bundle for similar crosstalk induced delay

Interconnect lengths (μm)	Equivalent area of bundled SWNTs for similar crosstalk induced delay		
	Area of bundled SWNT in nm^2 for equivalent MWNT area = 12.95 nm^2	Area of bundled SWNT in nm^2 for equivalent MWNT area = 43.71 nm^2	Area of bundled SWNT in nm^2 for equivalent MWNT area = 92.62 nm^2
100	512	968	2,048
200	578	1,152	2,450
500	1,352	2,450	4,418
800	1,458	4,232	6,728
1,000	1,568	4,608	8,712

Table 4.4 Percentage reduction in crosstalk induced delay for MWNTs with respect to SWNT bundle at equivalent areas

Corresponding number of shells in MWNTs	Percentage improvement in crosstalk induced delay for MWNTs with respect to bundled SWNTs at different interconnect lengths				
	100 μm	200 μm	500 μm	800 μm	1000 μm
10	33.2	45.6	65.2	71.7	72.8
20	30.9	42.2	56.4	66.2	69.3
30	26.4	35.1	45.1	59.5	66.6

Table 4.5 Percentage improvement in areas of MWNTs with respect to equivalent bundled SWNTs for same crosstalk induced delay

Corresponding area of MWNTs (nm^2)	Percentage improvement in area for MWNTs with respect to bundled SWNTs at interconnect lengths				
	100 μm	200 μm	500 μm	800 μm	1000 μm
12.95	97.4	97.7	99.0	99.1	99.2
43.71	95.4	96.2	98.2	98.9	99.0
92.62	95.4	96.2	97.9	98.6	98.9

In a denser bundle, the diameters of SWNTs is kept small that actually increases the number of SWNTs in that bundle. The small diameter SWNTs substantially reduces the coupling capacitance between the aggressor and victim lines and thereby, the overall crosstalk delay. Similarly, for more number of shells in MWNT, the diameter of outermost shell increases. This outermost shell carries lesser current than a smaller shell diameter that helps in reducing the coupling capacitance, thereby, reducing the crosstalk induced delay (shown in Table 4.2). However, from Table 4.2, it is observed that the same crosstalk induced delay is obtained for fewer number of shells using MWNTs compared to the equivalent number of SWNTs in bundle that encouragingly results in the reduction of area using MWNT interconnects as presented in Table 4.3. Furthermore, Table 4.4 presents that the percentage reduction in crosstalk induced delay substantially increases for longer interconnects, and on an average, this reduction is 52.4 % for MWNT compared to equivalent number of SWNTs in bundle. Consequently, an MWNT requires 97.8 % lesser area with respect to bundled SWNT interconnects (Table 4.5) for almost similar delay performance (Majumder et al. 2012a).

4.5 Crosstalk Induced Delay of Bundled SWNT, Bundled DWNT, and Single MWNT Interconnects

This section analyzes crosstalk induced delay at different global interconnect lengths for bundled SWNT/DWNT and single MWNT at 22 nm and 32 nm technology nodes as per the information provided by ITRS (2010). For global signals, such as clock line, the interconnect length can reach up to several millimeters. Dynamic crosstalk induced delay is analyzed for global interconnect lengths ranging from 800 μm to 2 mm when the victim line is switched in opposite direction with respect to the aggressors. Figures 4.14, 4.15 and 4.16 demonstrate the variation of crosstalk induced delay for different CNT structures at 22 nm and 32 nm technology nodes. It has been observed that the crosstalk induced delay reduces for lower technology node for different CNT structures. The reason behind this is that

Fig. 4.14 Crosstalk induced delay of SWNT bundle for different interconnect lengths at 22 nm and 32 nm technology nodes

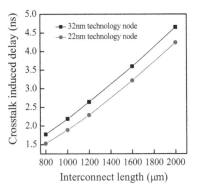

Fig. 4.15 Crosstalk induced delay of DWNT bundle for different interconnect lengths at 22 nm and 32 nm technology nodes

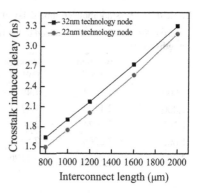

Fig. 4.16 Crosstalk induced delay of MWNT for different interconnect lengths at 22 nm and 32 nm technology nodes

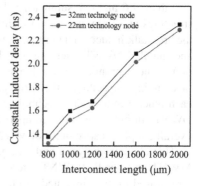

Table 4.6 Crosstalk induced delay at 32 nm technology node

Interconnect lengths (μm)	Crosstalk induced delay (in ns) for			Percentage reduction for single MWNT with respect to	
	Bundled SWNT	Bundled DWNT	Single MWNT	Bundled SWNT	Bundled DWNT
800	2.0011	1.6396	1.3777	31.15	15.97
1,000	2.3925	1.9046	1.5977	33.22	16.11
1,200	2.8063	2.1725	1.6818	40.07	22.58
1,600	3.6958	2.7259	2.0910	43.42	23.29
2,000	4.6674	3.2954	2.3423	49.81	28.92

the propagation delay under the effect of dynamic crosstalk is mainly influenced by the capacitive parasitic. As the technology scales down, the feature size decreases that in turn reduces the bundle width and height resulting in lesser number of CNTs in bundled SWNT/DWNT and lesser number of shells in MWNT. The coupling capacitance mainly depends on the diameter of outermost shell in MWNT and number of peripheral SWNTs/DWNTs in bundle. Thus, the crosstalk induced delay drastically reduces for lesser number of SWNTs/DWNTs in bundle and shells in MWNTs at global interconnect lengths.

Table 4.7 Crosstalk induced time delay at 22 nm technology node

Interconnect lengths (μm)	Crosstalk induced delay (in ns) for			Percentage reduction for single MWNT with respect to	
	Bundled SWNT	Bundled DWNT	Single MWNT	Bundled SWNT	Bundled DWNT
800	1.7654	1.4955	1.3225	25.08	11.56
1,000	2.1123	1.7474	1.5203	28.02	12.99
1,200	2.4924	2.0064	1.6252	34.79	18.99
1,600	3.3366	2.5674	2.0198	39.46	21.32
2,000	4.2875	3.1796	2.2960	46.44	28.63

Tables 4.6 and 4.7 summarize the crosstalk induced delay for different CNT structures at 32 nm and 22 nm technology nodes, respectively. It is observed that the crosstalk induced delay reduces for MWNT with respect to the bundled SWNT and bundled DWNT interconnects. Irrespective of technology nodes, this reduction is more pronounced for longer interconnects. It is due to the smaller value of coupling capacitance that primarily depends on the number of CNTs facing each other. For a specific technology node, the number of peripheral SWNTs and DWNTs in bundle is more compared to the MWNTs. It substantially reduces the overall coupling capacitance of MWNT and thereby, the crosstalk induced delay.

This chapter analyzed and compared the dynamic crosstalk induced delay of MWNT and bundled SWNT interconnects using three coupled line bus architecture. Based on the simulated results, equivalent number of SWNTs in bundle has been obtained for a fixed number of shells in MWNTs at different global interconnect lengths. It has been observed that the overall improvement in crosstalk induced delay and area are 52.4 % and 97.8 %, respectively, for MWNT in comparison to bundled SWNT interconnects. Therefore, MWNTs can be predicted as more appropriate candidate for future global VLSI interconnects.

Chapter 5
Mixed Carbon Nanotube Bundle

Abstract This chapter presents the modeling and performance analysis of novel mixed carbon nanotube bundle (MCB) interconnects. A MCB primarily consists of SWNTs and MWNTs of different diameters. This chapter introduces four different MCB arrangements based on the locations of SWNTs and MWNTs in densely packed bundle. The equivalent single conductor (ESC) model of MCB consists of one ESC model of bundled SWNT and the other ESC model of bundled MWNT interconnects. Using different MCB topologies, propagation delay, power dissipation, and crosstalk induced delay are analyzed for different interconnect lengths.

Keywords Mixed carbon nanotube bundle (MCB) · Equivalent single conductor (ESC) model · Power dissipation · In phase and out phase delays

5.1 Introduction

Several analyses have already been carried out based on the modeling and characterization of different single and bundled CNT structures. Banerjee and Srivastava (2006b) demonstrated that bundled SWNT is better for global and intermediate interconnects as long as their mean free path of 1 μm is maintained. However, Raychowdhury and Roy (2006) explained that MWNTs are more promising than bundled SWNT for on-chip interconnect applications. Recent analysis focuses on modeling and simulation of different arrangements of mixed CNT bundle (MCB) interconnects. The arrangement of MCB is more complicated than that of SWNTs or MWNTs. It is not trivial to analyze the MCBs using conventional modeling approach. Therefore, a new hierarchical model has been developed (Sathyakam and Mallick 2011; Majumder et al. 2012c) considering two types of CNTs, wherein one of them is SWNT and the other one is MWNT. Equivalent single conductor (ESC) models are developed for the bundled SWNT and bundled MWNT interconnects. These equivalent models are combined to build up an equivalent *RLC* model of proposed MCB arrangement. This chapter introduces the spatial arrangements of MCBs based on the location of SWNTs and MWNTs in the bundle.

© The Author(s) 2015
B.K. Kaushik and M.K. Majumder, *Carbon Nanotube Based VLSI Interconnects*, SpringerBriefs in Applied Sciences and Technology, DOI 10.1007/978-81-322-2047-3_5

5.2 Proposed MCB Topologies

Several researchers have already reported that conductivity of SWNTs is more than that of MWNTs since only the outermost shell of MWNT is properly contacted to the metal. Therefore, an MWNT exhibits poor conduction with respect to SWNT (Collins et al. 2001b). It has been reported that the outermost shell of MWNT exhibits not only even/odd multiples of conductance but also the fractional and non-integer quantum conductance because of inter-wall interactions (Sanvito et al. 2000; Subash et al. 2013). Depending on these facts, four different MCB topologies (such as MCB-I, MCB-II, MCB-III, and MCB-IV) are introduced that contains spatial arrangements of SWNTs and MWNTs in a bundle. In MCB-I, the strongly conducting SWNTs are placed at the center of the bundle, while the insignificantly conducting MWNTs are

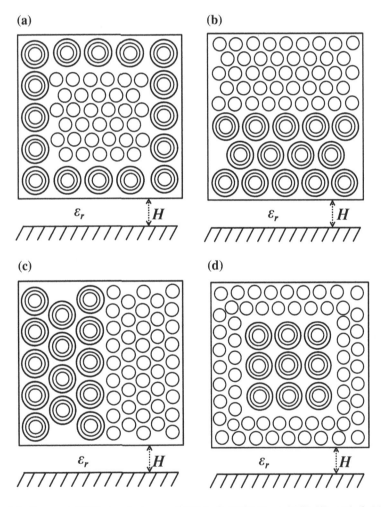

Fig. 5.1 Proposed MCB topologies: **a** MCB-I, **b** MCB-II, **c** MCB-III, and **d** MCB-IV (Majumder et al. 2012c)

located on the periphery as shown in Fig. 5.1a. SWNTs and MWNTs occupy equal halves as horizontal and vertical arrangements in MCB-II and MCB-III, respectively, as presented in Figs. 5.1b, c, respectively. Finally, Fig. 5.1d shows the MCB-IV arrangement wherein SWNTs are placed at periphery to the centrally located MWNTs.

5.3 ESC Model of MCB Interconnects

Based on the arrangements (Fig. 5.1), this section presents an equivalent *RLC* model of MCB. The equivalent model is the combination of the ESC models of bundled SWNT and bundled MWNT. ESC models of bundled CNTs are developed using the transmission line theory (Amore et al. 2010). The general ESC circuit model of a multi-conductor nanoline is shown in Fig. 5.2, wherein effective per unit length (*p.u.l.*) total inductance (l_{ESC}) and capacitance (c_{ESC}) can be expressed as

$$l_{ESC} = l_{k\text{-}ESC} + l_{e\text{-}ESC} \tag{5.1}$$

$$c_{ESC} = \left(\frac{1}{c_{q\text{-}ESC}} + \frac{1}{c_{e\text{-}ESC}} \right)^{-1} \tag{5.2}$$

For bundled SWNT and MWNT, the effective *p.u.l.* kinetic inductance ($l_{k\text{-}ESC}$) and quantum capacitance ($c_{q\text{-}ESC}$) are expressed as (5.3) and (5.4), respectively, where n_{CNT} is the number of SWNTs and MWNTs in the MCB.

$$l_{k\text{-}ESC} = \frac{l_{k0}}{2n_{CNT}} \quad \text{where } l_{k0} = \frac{h}{2e^2 v_F} \approx 16.1 \text{ mH/m} \tag{5.3}$$

$$c_{q\text{-}ESC} = 2n_{CNT}c_{q0} \quad \text{where } c_{q0} = \frac{2e^2}{h v_F} \approx 96.8 \text{ pF/m} \tag{5.4}$$

where l_{k0} and c_{q0} are the *p.u.l.* kinetic inductance and quantum capacitance, respectively (Jorio et al. 2008; Ji et al. 2006).

At both ends, the tube is terminated by an equivalent lumped resistance R_{tESC} as shown in Fig. 5.3. The R_{tESC} can be expressed as (Pandya et al. 2012)

$$R_{tESC} = \frac{R_0}{2n_{CNT}} + R_{mc} \tag{5.5}$$

where R_0 is the intrinsic *dc* resistance of a quantum wire (i.e., CNT) and is equal to $h/4e^2 \approx 6.45 \text{ k}\Omega$. The R_{mc} represents imperfect metal-nanotube contact

Fig. 5.2 ESC of bundled SWNT or bundled MWNT

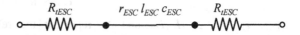

R_{tESC} $r_{ESC} \, l_{ESC} \, c_{ESC}$ R_{tESC}

$Z{=}0$ $Z{=}l$

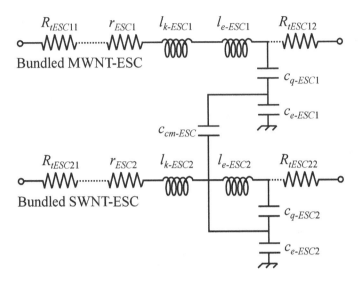

Fig. 5.3 Equivalent electrical model of MCB interconnects

resistance that exhibits a typical value of 3.5 kΩ depending on the fabrication process (Ji et al. 2006; Avouris et al. 2003). The scattering resistance (r_{ESC}) primarily appears due to the static impurity scattering, defects, line edge roughness scattering (LER), and acoustic phonon scattering. The *p.u.l.* r_{ESC} can be expressed as

$$r_{\text{ESC}} = \frac{h}{4e^2 \lambda_{\text{mfp}}} \tag{5.6}$$

where h is known as Planck's constant and λ_{mfp} is the mean free path. The effective *p.u.l.* external capacitance ($c_{e\text{-}ESC}$) is the electrostatic capacitance of the SWNTs and/or MWNTs facing the ground plane and can be expressed as (5.7). On the other hand, the effective *p.u.l.* magnetic inductance ($l_{e\text{-}ESC}$) is the stored energy for a given amount of current flow and can be expressed as (5.8).

$$c_{e\text{-ESC}} = \frac{2\pi \varepsilon_0 \varepsilon_r}{\ln \left(H/D_y \right)} \times N_y \tag{5.7}$$

$$l_{e\text{-ESC}} = \frac{\mu_0 \varepsilon_0}{c_{e-\text{ESC0}}} \tag{5.8}$$

where $\varepsilon_r \approx 2.2$, H, D_y, and N_y represent the relative permittivity of the medium, distance of the MCB from ground plane, diameter and number of SWNTs/MWNTs facing the ground plane, respectively. $c_{e\text{-}ESC0}$ is the *p.u.l.* electrostatic capacitance embedded in free space (Sarto et al. 2010).

Using transmission line model parameters, the electrical equivalent model of MCB is shown in Fig. 5.3. It is assumed that all shells of MWNTs and all SWNTs in bundle are parallel. Tubes within the same or different bundles experience a capacitance, known as coupling capacitance ($c_{cm\text{-}ESC}$) and can be expressed as (Subash et al. 2013)

$$c_{cm-ESC} = \pi \varepsilon_0 \left/ \ln\left[\left(\frac{S_{C-C}}{D_{mean}}\right) + \left(\sqrt{\left(\frac{S_{C-C}}{D_{mean}}\right)^2 + 1}\right) \right] \right. \tag{5.9}$$

where S_{C-C} is the distance between the center of any two CNTs and D_{mean} represents mean diameter of adjacent SWNTs and/or MWNTs in the bundle. While considering a parallel layout of CNT interconnects, the $c_{cm\text{-}ESC}$ plays in significant role than the quantum and electrostatic capacitances.

5.4 Performance Analysis of MCB Based Interconnects

This section analyzes the power, delay, and crosstalk performance of different MCB topologies based on the equivalent electrical circuit model.

5.4.1 Propagation Delay and Power Dissipation of MCB Topologies

Propagation delay and power dissipation of MCBs are analyzed for different interconnect lengths ranging from 100 μm to 1,000 μm using the geometries suggested in Fig. 5.1a, d. Simulation setup uses CMOS driver at 32 nm technology node wherein the technology parameters (length and width) for NMOS are taken as 32 nm and 640 nm while for PMOS, these values are 32 nm and 1,280 nm, respectively. As shown in Fig. 5.4, the input rise time is triggered at 90 %, whereas the output fall time is target at 10 %. The propagation delay is analyzed for MCB-I and MCB-IV with 20 distributed segments at different global interconnect lengths. For simulation purpose, the diameter of each SWNT is taken as 1 nm, whereas the outer

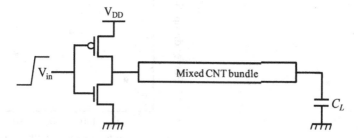

Fig. 5.4 Driver-interconnect-load system employing CMOS driver

shell diameter for MWNT is considered as 5 nm, such that the metallicity property does not change. A driver-interconnect-load (DIL) system with CMOS driver, is terminated with an output load capacitance (C_L) of 10 fF (Pandya et al. 2012).

Figures 5.5 and 5.6 demonstrate the comparison of propagation delay and power dissipation of two different MCB topologies (MCB-I and MCB-IV). It has been observed that propagation delay in MCB-I is more compared to MCB-IV at global interconnect lengths. The reason behind this is the structural difference between different MCBs. From Fig. 5.1, it is observed that MWNTs are at periphery in MCB-I, whereas for MCB-IV, the MWNTs are located at the center. Therefore, the number of MWNTs in MCB-I is more compared to MCB-IV. As the conductivity of MWNTs is smaller in comparison to the SWNTs, the propagation delay is more for MCB-I in comparison to MCB-IV. Due to higher current, the power dissipation of MCB-IV is more in comparison to MCB-I for different interconnect lengths.

Fig. 5.5 Propagation delay of MCB-I and MCB-IV for different interconnect lengths

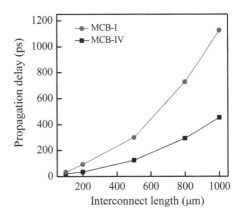

Fig. 5.6 Power dissipation of MCB-I and MCB-IV for different interconnect lengths

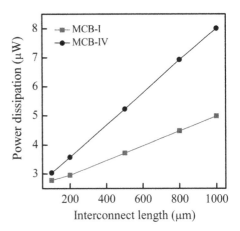

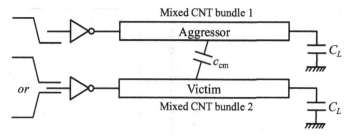

Fig. 5.7 Capacitively coupled interconnect lines

5.4.2   Crosstalk Induced Delay of MCB Topologies

Propagation delay under the influence of dynamic crosstalk has been analyzed using capacitively coupled interconnect lines as shown in Fig. 5.7. Out of these two lines, one is referred as aggressor, and the other one as victim (Kreupl et al. 2002). Interconnect line in the bus architecture is represented by the ESC model of MCB (Fig. 5.3). In the bus architecture of Fig. 5.7, CMOS driver is used instead of resistive driver for accurate estimation of crosstalk induced delay (Majumder et al. 2012c). To analyze the crosstalk induced delay, internal coupling capacitance (c_{cm}) is considered between aggressor and victim lines. The c_{cm} primarily depends on the spacing between aggressor and victim. The coupled interconnect lines have load capacitance ($C_L = 10aF$) and power supply voltage ($V_{dd} = 1V$) (Majumder et al. 2012c). Crosstalk induced delay is observed for different global interconnect lengths ranging from 100 μm to 1,000 μm.

Crosstalk induced delay is analyzed for MCB-I, MCB-II, MCB-III, and MCB-IV topologies. Using capacitively coupled interconnect lines, the in phase and out phase delays are observed at victim line with respect to the aggressor. It has been observed that propagation delay under the influence of crosstalk increases for longer interconnects. Figures 5.8, 5.9, 5.10 and 5.11 demonstrate the variation of crosstalk induced delay of mixed CNT bundles for different interconnect lengths. It is observed that the in phase delay is lesser compared to the out phase delay for fixed interconnect length. The reason behind this is the Miller capacitive effect that leads to almost doubling of c_{cm}. Under out phase transitions, the Miller coupling factor (MCF) tends to a value of 2. This fact can be explained by the following expression of crosstalk induced delay (Rabaey 2002)

$$t = gC_W(0.38R_W + 0.69R_D) \tag{5.10}$$

where C_W and R_W represent the capacitance to ground and resistance of the wire, while R_D is equivalent resistance of the driver. The correction factor g introduces the crosstalk effect and is a function of the ratio $r = C_i/C_W$, where C_i is the inter-wire capacitance. When the two wires (aggressor and victim) makes transition in the same direction, the C_i has no effect, and therefore, $g = 1$. The worst case occurs for opposite transition in aggressor and victim that leads to factor $g = 1 + r$. Therefore, it can be concluded that the out phase delay is more compared to the in phase delay.

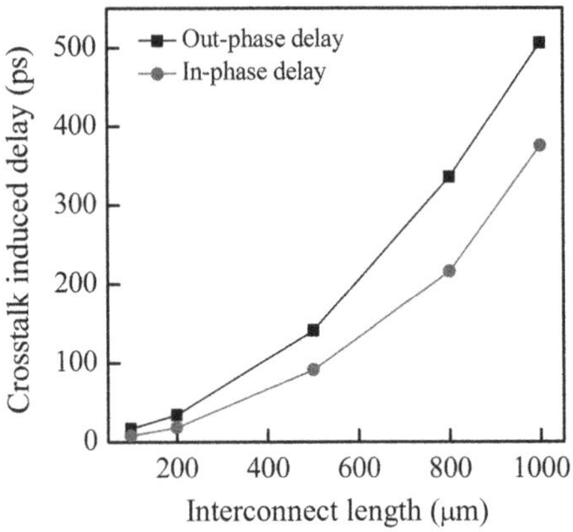

Fig. 5.8 Crosstalk induced delay of MCB-I for different interconnect lengths

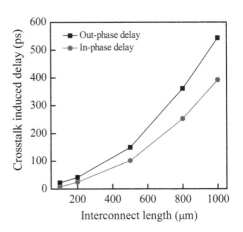

Fig. 5.9 Crosstalk induced delay of MCB-II for different interconnect lengths

Fig. 5.10 Crosstalk induced delay of MCB-III for different interconnect lengths

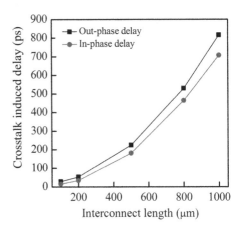

Fig. 5.11 Crosstalk induced delay of MCB-IV for different interconnect lengths

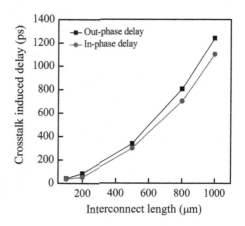

Table 5.1 Percentage reduction in crosstalk induced delay of MCB-I compared to other MCBs

Interconnect lengths (μm)	Percentage reduction in crosstalk induced delay of MCB-I w.r.t.					
	MCB-II		MCB-III		MCB-IV	
	In phase	Out phase	In phase	Out phase	In phase	Out phase
100	7.01	4.85	39.13	37.50	54.56	51.72
200	8.91	6.75	47.31	36.74	64.29	53.73
500	12.45	7.79	48.71	38.16	66.31	57.09
800	16.67	16.22	49.73	37.56	68.94	58.47
1,000	23.07	22.22	52.71	36.95	69.74	58.86

From Figs. 5.8, 5.9, 5.10 and 5.11, it can be inferred that the in phase and out phase delays reduces for MCB-I compared to MCB-II, MCB-III, and MCB-IV. The percentage reduction in phase and out phase delays for MCB-I is summarized in Table 5.1. It is observed that the overall reduction for in phase and out phase delays using MCB-I are 64.77 % and 55.97 %, respectively, compared to MCB-IV. It is due to the spatial arrangement of MCB-I wherein MWNTs are placed at periphery. The MWNTs in periphery primarily serve as shield between two coupled lines. Due to this shielding, the inter-CNT capacitance (c_{cm}) has much lower effect on MCB-I with respect to other MCBs. Therefore, crosstalk induced delay of MCB-I drastically reduces both for the in phase and out phase switching scenario.

This chapter presented a hierarchical modeling approach for MCB interconnect that is a combination of ESC models of bundled SWNT and MWNT. Depending on the location of SWNTs and MWNTs, different MCB structures are introduced. Propagation delay under the influence of dynamic crosstalk is demonstrated at victim line both for in phase and out phase switching. It has been observed that crosstalk induced delay is minimum for MCB structure that has MWNTs at peripheral and SWNTs at centre in the bundle because of the shielding effect of MWNTs.

References

Agarwal D (2002) Optical interconnects to silicon chips using short pulses. PhD thesis, Stanford University, Stanford

Agarwal K, Sylvester D, Blaauw D (2006) Modeling and analysis of crosstalk noise in coupled *RLC* interconnects. IEEE Trans Comput Aided Des Integr Circuits Syst 25(5):892–901

Agrawal S, Raghuveer MS, Ramprasad R, Ramanath G (2007) Multishell carrier transport in multiwalled carbon nanotubes. IEEE Trans Nanotechnol 6(6):722–726

Ago H, Petritsch K, Shaffer MSP, Windle AH, Friend RH (1999) Composites of carbon nanotubes and conjugated polymers for photovoltaic devices. Adv Mater 11(15):1281–1285

Ahmed H (2010) A new method for fabrication and laser treatment of nano composites. J Am Sci 6(7):149–154

Amore MD, Sarto MS, Tamburrano A (2010) Fast transient analysis of next-generation interconnects based on carbon nanotubes. IEEE Trans Electromagn Compat 52(2):496–503

Appenzeller J, Lin Y-M, Knotch J, Zhihong C, Avouris P (2005) Comparing carbon nanotube transistors—the ideal choice: a novel tunneling device design. IEEE Trans Electron Devices 52(12):2568–2576

Arbiol J, Kalache B, Roca P, Morante JR, Morral AF (2007) Influence of Cu as a catalyst on the properties of silicon nanowires synthesized by the vapour–solid–solid mechanism. Nanotechnology 18(30):1–8

Avouris P, Appenzeller J, Martel R, Wind SJ (2003) Carbon nanotube electronics. Proc of IEEE 91(11):1772–1784

Avorious P, Chen Z, Perebeions V (2007) Carbon-based electronics. Nat Nanotechnol 2(10):605–613

Awschalom DD, Flatte ME (2007) Challenges for semiconductor spintronics. Nat Phys 3:153–159

Banerjee K, Srivastava N (2006a) Are carbon nanotubes the future of VLSI interconnections? In: Proceedings of ACM/IEEE design automation conference, San Francisco, CA, USA, pp 809–814

Banerjee K, Srivastava N (2006b) Performance analysis of carbon nanotube interconnects for VLSI application. In: Proceedings of 43rd ACM/IEEE design automation conference, pp 809–814

Bellucci S, Onorato P (2010) The role of the geometry in multiwall carbon nanotube interconnects. J Appl Phys 108(7):073704-1–073704-9

Berber S, Kwon Y-K, Tomanek D (2000) Unusually high thermal conductivity of carbon nanotubes. Phys Rev Lett 84(20):4613–4616

Bhattacharya S, Mahapatra S (2010) Negative differential conductance and effective electron mass in highly asymmetric ballistic bilayer graphene nanoribbon. Phys Lett A 374(28):2850–2855

© The Author(s) 2015
B.K. Kaushik and M.K. Majumder, *Carbon Nanotube Based VLSI
Interconnects*, SpringerBriefs in Applied Sciences and Technology,
DOI 10.1007/978-81-322-2047-3

Borkar S, Kamik T, Narendra S, Tschanz J, Keshavarzi A, De V (2003) Parameter variations and impact on circuits and microarchitecture. In: Proceedings of IEEE Design automation conference (DAC), pp 338–342

Bourianoff G (2003) The future of nanocomputing. Computer 36(8):44–53

Bourlon B, Miko C, Forro L, Glattli DC, Bachtold A (2004) Determination of the intershell conductance in multiwalled carbon nanotubes. Phys Rev Lett 93(17):176806-1–176806-4

Burke PJ (2002) Lüttinger liquid theory as a model of the gigahertz electrical properties of carbon nanotubes. IEEE Trans Nanotechnol 1(3):129–144

Cavin RV, Zhirnov VV, Herr DJC, Avila A, Hutchby J (2006) Research directions and challenges in nanoelectronics. J Nanopart Res 8(6):841–858

Che G, Lakshmi BB, Martin CR, Fisher ER (1998) Chemical vapor deposition based synthesis of carbon nanotubes and nanofibers using a template method. Chem Mater 10(1):260–267

Chen G, Chen H, Haurylau M, Nelson NA, Fauchet PM, Friedman EG, Albonesi DH (2005) Predictions of CMOS compatible on-chip optical interconnect. In: Proceedings of ACM 2005 international workshop on system level interconnect prediction (SLIP-05), San Francisco, CA, USA, pp 12–20

Chen WC, Yin WY, Jia L, Liu QH (2009) Electrothermal characterization of single-walled carbon nanotube (SWCNT) interconnect arrays. IEEE Trans Nanotechnol 8(6):718–728

Chibante LPF, Thess A, Alford JM, Diener MD, Smalley RE (1993) Solar generation of the fullerenes. J Phys Chem 97(34):8696–8700

Cho H, Koo KH, Kapur P, Saraswat KC (2008) Performance comparisons between Cu/low-κ, carbon-nanotube, and optics for future on-chip interconnects. IEEE Trans Electron Devices 29(1):122–124

Choi WB, Chung DS, Kang JH, Kim HY, Jin YW, Han IT, Lee YH, Jung JE, Lee NS, Park GS, Kim JM (1999) Fully sealed, high-brightness carbon-nanotube field-emission display. Appl Phys Lett 75(20):3129–3131

Close GF, Wong HSP (2008) Assembly and electrical characterization of multiwall carbon nanotube interconnects. IEEE Trans Nanotechnol 7(5):596–600

Collins G, Hersam M, Arnold M, Martel R, Avouris P (2001a) Current saturation and electrical breakdown in multiwalled carbon nanotubes. Phys Rev Lett 86(14):3128–3131

Collins PG, Arnold MS, Avouris P (2001b) Engineering carbon nanotubes and nanotube circuits using electrical breakdown. Science 292(5517):706–709

Dadgour H, Cassell AM, Banerjee K (2008) Scaling and variability analysis of CNT-based NEMS devices and circuits with implications for process design. In: Proceedings of IEEE international electron device meeting (IEDM), San Francisco, CA, USA, pp 529–532

Dannberg P, Erdmann L, Krehl A, WaEchter C, BraEuer A (2000) Integration of optical interconnects and optoelectronic elements on wafer-scale. Mater Sci Semicond Process 3(5–6):437–441

David PS (2010) Plasmonics integrated optics going the last few microns. In: Proceedings of IEEE photonics society summer topical meeting series, USA, pp 72–73

Dresselhaus M, Dresselhaus G, Avouris P (2001) Carbon nanotubes: Synthesis, structure, properties and applications. Topics in applied research 80, Springer

Du C, Yeh J, Pan N (2005) High power density supercapacitors using locally aligned carbon nanotube electrodes. Nanotechnology 16(4):350–353

Eachempati S, Vijaykrishnan N, Nieuwoudt A, Massoud Y (2009) Predicting the performance and reliability of future field programmable gate arrays routing architectures with carbon nanotube bundle interconnect. IET Circuits Devices Syst 3(2):64–75

Ebbesen TW (1996) Carbon nanotubes: Preparation and properties. CRC Press

Echtermeyer TJ, Lemme MC, Baus M, Szafranek BN, Geim AK, Kurz H (2008) Nonvolatile switching in graphene field-effect devices. IEEE Electron Device Lett 29(8):952–954

Elgamel MA, Bayoumi MA (2003) Interconnect noise analysis and optimization in deep submicron technology. IEEE Circuits Syst Mag 3(4):6–17

Fathi D, Forouzandeh B, Mohajerzadeh S, Sarvari R (2009) Accurate analysis of carbon nanotube interconnects using transmission line model. Micro Nano Lett 4(2):116–121

Forro L, Salvetat JP, Bonard JM, Bacsa R, Thomson NH, Garaj S, Thien-Nga L, Gaál R, Kulik A, Ruzicka B, Degiorgi L, Bachtold A, Schönenberger C, Pekker S, Hernadi K (2002) Electronic and mechanical properties of carbon nanotubes. Fundamental materials research, Springer, pp 297–320

Galatsis K, Khitun A, Ostroumov R, Wang KL, Dichtel WR, Plummer E, Stoddart JF, Zink JI, Lee JY, Xie Y-H, Kim K W (2009) Alternate state variables for emerging nanoelectronic devices. IEEE Trans Nanotechnol 8(1):66–75

Gengchiau L, Neophytos N, Nikonov DE, Lundstrom MS (2007) Performance projections for ballistic graphene nanoribbon field-effect transistors. IEEE Trans Electron Devices 54(4):677–682

Giustiniani A, Tucci V, Zamboni W (2010) Modeling issues and performance analysis of high-speed interconnects based on a bundle of SWNT. IEEE Trans Electron Devices 57(8):1978–1986

Glenn Research Center (2002) 2002 Research and technology. National Aeronautics and Space Administration (NASA), NASA/TM-2003-211990, Online available at: http://books.google.co.in/books?isbn=1428918205

Goel AK (2007) High-speed VLSI interconnections, 2nd edn. Wiley-IEEE Press, New York

Geol AK (2008) Nanotube and other interconnects for nanotechnology circuits. In: Proceedings of IEEE Canadian conference on electrical and computer engineering (CCECE 2008), pp 000189–000192

Gruner G (2005) Carbon nanotube transistors for biosensing applications. Anal Bioanal Chem 384(2):322–335

Gunlycke D, Lawler HM, White CT (2007) Room-temperature ballistic transport in narrow graphene strips. Phys Rev B 75(8):085418-1–085418-5

Haruehanroengra S, Wang W (2007) Analyzing conductance of mixed carbon nanotube bundles for interconnect applications. IEEE Electron Device Lett 28(8):756–759

Harris PJF (1999) Carbon nanotubes and related structures: New materials for the twenty-first century. Cambridge University Press, p 49

Hasan S, Salahuddin S, Vaidyanathan M, Alam MA (2006) High-frequency performance projections for ballistic carbon-nanotube transistors. IEEE Trans Nanotechnol 5(1):14–22

Hsieh JY, Lu JM, Huang MY, Hwang CC (2006) Theoretical variations in the Young's modulus of single-walled carbon nanotubes with tube radius and temperature: A molecular dynamics study. Nanotechnol 17:3920–3924

Huang Y, Yin WY, Liu QH (2008) Performance prediction of carbon nanotube bundle dipole antennas. IEEE Trans Nanotechnol 7(3):331–337

Hu C, Liu C, Chen L, Meng C, Fan S (2010) A demo opto-electronic power source based on single-walled carbon nanotube sheets. ACS Nano 4(8):4701–4706

Hyperion Catalysis [Online] Available: http://www.fibrils.com

International Technology Roadmap for Semiconductors (ITRS). Online Availablehttp://www.itrs.net

Jamal O, Naeemi A (2011) Ultralow-power single-wall carbon nanotube interconnects for sub-threshold circuits. Res Lett IEEE Trans Nanotechnol 10(1):99–101

Journet C, Bernier P (1998) Production of carbon nanotubes. Appl Phys A, Springer 67(1):1–9

Jousseaume V, Renard VT (2010) Cu based catalysts can make CMOS compatible Si nanowires: toward reconfigurable interconnects. In: Proceedings of IEEE international conference on interconnect technology (IITC 2010), France, pp 1–3, June 2010

Javey A, Kong J (2009) Carbon nanotube electronics. Springer, Berlin

Ji Y, Lin YJ, Wong JSC (2006) Buckypaper's fabrication and application to passive vibration control. In: Proceedings of 1st IEEE international conference on nano/micro engineered and molecular systems (NEMS '06), pp 725–729

Jorio A, Dresselhaus G, Dresselhaus MS (2008) Carbon nanotubes: advanced topics in the synthesis, structure, properties and applications. Springer, Berlin

Jornet JM, Akyildiz IF (2010) Graphene-based nano-antennas for electromagnetic nano communications in the terahertz band. In: Proceedings of IEEE 4th European conference on antennas and propagations (EuCAP 2010), pp 1–5

Kapur P, Saraswat KC (2002) Optical interconnects for future high performance integrated circuits. Physica E Low-Dimens Syst Nanostruct 16(3/4):620–627

Kase D, Shiba K, Zhu J, Kasuya D, Yudasaka M, Iijima S (2003) Toward development of nano-materials composed of artificial proteins and nano-carbons. In: Proceedings of 3rd IEEE international conference on nanotechnology 2003 (IEEE-NANO 2003), vol 1, pp 386–389

Kaushik BK, Goel S, Rauthan G (2007) Future VLSI interconnects: optical fiber or carbon nanotube—a review. Microelctron Int 24(2):53–63

Kaushik BK, Sarkar S (2008) Crosstalk analysis for a CMOS-gate-driven coupled interconnects. IEEE Trans Comput Aided Des Integr Circuits Syst 27(6):1150–1154

Kaushik BK, Sankar S, Agarwal RP, Joshi RC (2010) An analytical approach to dynamic crosstalk in coupled interconnects. Microelectron J 41(2–3):85–92

Kim W, Javey A, Tu R, Cao J, Wang Q, Dai H (2005) Electrical contacts to carbon nanotubes down to 1 nm in diameter. Appl Phys Lett 87(17):173101

Koo K-H, Cho H, Kapur P, Saraswat KC (2007) Performance comparison between carbon nanotubes, optical and Cu for future high- performance on-chip interconnect applications. IEEE Trans Electron Devices 54(12):3206–3215

Koo K-H, Kapur P, Saraswat KC (2009) Compact performance models and comparisons for gigascale on-chip global interconnect technologies. IEEE Trans Electron Devices 56(9):1787–1798

Kreupl F, Graham AP, Deusberg GS, Steinhogl W, Liebau M, Unger E, Honlein W (2002) Carbon nanotubes in interconnect applications. Microelectron Eng 64(1–4):399–408

Kshirsagar C, Li H, Kopley TE, Banerjee K (2008) Accurate intrinsic gate capacitance model for carbon nanotube-array based FETs considering screening effect. IEEE Electron Device Lett 29(12):1408–1411

Laplazeb D, Bernierb P, Journetb C, Vié V, Flamant G, Lebrun M (1997) Carbon sublimation using a solar furnace. Synth Met 86(1-3):2295–2296

Laval S (2000) Optical interconnects: the challenge. Comptes Rendus de l'Académie des Sciences-Series IV, Paris, pp 941–949

Lemme MC, Echtermeyer TJ, Baus M, Kurz H (2007) A graphene field-effect device. IEEE Electron Device Lett 28(4):282–284

Li H, Banerjee K (2008a) High-frequency effects in carbon nanotube interconnects and implications for on-chip inductor design. In: Proceedings of IEEE international electron device meeting (IEDM 2008), San Francisco, CA, USA, pp 525–528

Li H, Yin WY, Banerjee K, Mao JF (2008b) Circuit modeling and performance analysis of multi-walled carbon nanotube interconnects. IEEE Trans Electron Devices 55(6):1328–1337

Li H, Xu C, Srivastava N, Banerjee K (2009a) Carbon nanomaterials for next-generation interconnects and passives: physics, status and prospects. IEEE Trans Electron Devices 56(9):1799–1821

Li H, Banerjee K (2009b) High-frequency analysis of carbon nanotube interconnects and implications for on-chip inductor design. IEEE Trans Electron Devices 56(10):2202–2214

Li H, Xu C, Banerjee K (2010) Carbon nanomaterials: the ideal interconnect technology for next-generation ICs. IEEE Des Test Comput 27(4):20–31

Li J, Ye Q, Cassell A, Ng HT, Stevens R, Han J, Meyyappan M (2003) Bottom-up approach for carbon nanotube interconnects. Appl Phys Lett 82(15):2491–2493

Lim SC, Jang JH, Bae DJ, Han GH, Lee S, Yeo I-S, Lee YH (2009) Contact resistance between metal and carbon nanotube interconnects: effect of work function and wettability. Appl Phys Lett 95(26):264103-1–264103-3

Lin W, Xiu Y, Zhu L, Moon KS, Wong CP (2008) Assembling of carbon nanotube structures by chemical anchoring for packaging applications. In: Proceedings of IEEE 58th electronic components and technology conference (ECTC 2008), pp 421–426

Lin Z-D, Hsiao C-H, Young S-J, Huang C-S, Chang S-J, Wang S-B (2013) Carbon nanotubes with adsorbed Au for sensing gas. IEEE Sens J 13(6):2423–2427

Lu F, Gu L, Meziani MJ, Wang X, Luo PG, Veca LM, Cao L, Sun Y-P (2009) Advances in bioapplications of carbon nanotubes. Adv Mater 21(2):139–152

Maffucci A, Miano G, Rubinacci G, Tamburrino A, Villone F (2008a) Plasmonics, CNT, conventional nano-interconnects: a comparison of propagation properties. In: Proceedings of IEEE workshop on signal propagation on interconnects (SPI), Avignon, pp 1–4

Maffucci A, Miano G, Villone F (2008b) Performance comparison between metallic carbon nanotube and copper nano-interconnects. IEEE Trans Adv Packag 31(4):692–699

Maffucci A, Miano G, Villone F (2009) A new circuit model for carbon nanotube interconnects with diameter-dependent parameters. IEEE Trans Nanotechnol 8(3):345–354

Majumder MK, Kaushik BK, Manhas SK (2011a) A comparative study of SWNT bundle and MWNT in terms of area and propagation delay for global interconnects. Int J Contemp Res Eng Technol 1(1):45–60

Majumder MK, Kaushik BK, Manhas SK (2011b) A comparative analysis of single walled CNT bundle and multi walled CNT as future global VLSI interconnects. Int J Comput Appl (IJCA) 2(6):32–38 (Special Issue on Evolution in Networks and Computer Communications)

Majumder MK, Kaushik BK, Manhas SK (2011c) Performance comparison between single wall carbon nanotube bundle and multiwall carbon nanotube for global interconnects. In: Proceedings of IEEE international conference on networks and computer communications (ETNCC 2011), Udaipur, Rajasthan, India, pp 104–109

Majumder MK, Kaushik BK, Manhas SK (2011d) Comparison of propagation delay characteristics for single-walled CNT bundle and multiwalled CNT in VLSI interconnects. In: Proceedings of IEEE international conference on recent advances in intelligent computational systems (RAICS 2011), Trivandum, India, pp 911–916

Majumder MK, Pandya ND, Kaushik BK, Manhas SK (2012a) Analysis of MWNT and bundled SWNT interconnects: impact on crosstalk and area. IEEE Electron Device Lett 33(8):1180–1182

Majumder MK, Pandya ND, Kaushik BK, Manhas SK (2012b) Analysis of crosstalk delay and area for MWNT and bundled SWNT for global VLSI interconnects. In: Proceedings of IEEE international symposium on quality electronic design 2012 (ISQED 2012), Santa Clara, CA, USA, pp 291–297

Majumder MK, Pandya ND, Kaushik BK, Manhas SK (2012c) Dynamic crosstalk effect in mixed CNT bundle interconnects. IET Electron Lett 48(7):384–385

Manney S, Nakhla MS, Zhang QJ (1992) Time domain analysis of non-uniform frequency dependent high-speed interconnects. In: Proceedings of IEEE/ACM international conference on computer-aided design (ICCAD '92), CA, USA, pp 449–453

Martin Y, Zhenyu L, Tsutsumi T, Shou R, Nakano M, Suehiro J, Ohtsuka S (2012) Detection of SF_6 decomposition products generated by DC corona discharge using a carbon nanotube gas sensor. IEEE Trans Dielectr Electr Insul 19(2):671–676

Mezhiba AV, Friedman EG (2002) Inductive properties of high-performance power distribution grids. IEEE Trans Very Large Scale Integr Syst (VLSI) 10(6):762–776

Mezhiba AV, Friedman EG (2004) Impedance characteristics of power distribution grids in nanoscale integrated circuits. IEEE Trans Very Large Scale Integr Syst (VLSI) 12(11):1148–1155

Miano G, Villone F (2005) An integral formulation for the electrodynamics of metallic carbon nanotubes based on a fluid model. IEEE Trans Antennas Propag 54(10):2713–2724

Minto E (2004) Tuning the band structure of carbon nanotubes. Ph.D. thesis, Cornell University

Misewich JA, Martel R, Avouris P, Tsang JC, Heinze S, Tersoff J (2003) Electrically induced optical emission from a carbon nanotube FET. Science 300(5620):783–786

Mizuno K, Ishii J, Kishida H, Hayamizu Y, Yasuda S, Futaba DN, Yumura M, Hata K (2009) A black body absorber from vertically aligned single-walled carbon nanotubes. In: Proceedings of National Academy of Sciences, pp 6044–6077

Naeemi A, Meindl JD (2005) Monolayer metallic nanotube interconnects: promising candidates for short local interconnects. IEEE Electron Device Lett 26(8):544–546

Naeemi A, Meindl JD (2006) Compact physical models for multiwall carbon-nanotube interconnects. IEEE Electron Device Lett 27(5):338–340

Naeemi A, Meindl JD (2007a) Physical modeling of temperature coefficient of resistance for single- and multi-wall carbon nanotube interconnects. IEEE Electron Device Letters 28(2):135–138

Naeemi A, Meindl JD (2007b) Design and performance modeling for single-walled carbon nanotubes as local, semiglobal, and global interconnects in gigascale integrated systems. IEEE Trans Electron Devices 54(1):26–37

Naeemi A, Meindl JD (2008) Performance modeling for single- and multiwall carbon nanotubes as signal and power interconnects in gigascale systems. IEEE Trans Electron Devices 55(10):2574–2582

Naeemi A, Sarvari R, Meindl JD (2005) Performance comparison between carbon nanotube and copper interconnects for gigascale integration (GSI). IEEE Electron Device Lett 26(2):84–86

Ni L, Demami F, Rogel R, Salaiin AC, Pichon L (2009) Fabrication and electrical characterization of silicon nanowires based resistors. Mater Sci Eng 6(1):1–4

Nieuwoudt A, Massoud Y (2006a) Evaluating the impact of resistance in carbon nanotube bundles for VLSI interconnect using diameter-dependent modeling techniques. IEEE Trans Electron Devices 53(10):2460–2466

Nieuwoudt A, Massoud Y (2006b) Understanding the impact of inductance in carbon nanotube bundles for VLSI interconnect using scalable modeling techniques. IEEE Trans Nanotechnol 5(6):758–765

Nieuwoudt A, Massoud Y (2007) Performance implications of inductive effects for carbon-nanotube bundle interconnect. IEEE Electron Device Lett 28(4):305–307

Nieuwoudt A, Massoud Y (2008) On the optimal design, performance, and reliability of future carbon nanotube-based interconnect solutions. IEEE Trans Electron Devices 55(8):2097–2110

Ngo Q, Petranovic D, Krishnan S, Cassell AM, Ye Q, Li J, Meyyappan M, Yang CY (2004) Electron transport through metal–multiwall carbon nanotube interfaces. IEEE Trans Nanotechnol 3(2):311–317

Oh S-Y, Jung W-Y, Kong J-T, Lee K-H (1999) Interconnect modeling in deep sub-micron design. In: Proceedings of IEEE 6th international conference on VLSI and CAD, pp 73–80

Ong YT, Ahmad AL, Zein SHS, Tan SH (2010) A review on carbon nanotubes in an environmental protection and green engineering perspective. Braz J Chem Eng 27(2):227–242

Ouyang Y, Yoon Y, Fodor JK, Guo J (2006) Comparison of performance limits for graphene nanoribbon and carbon nanotube transistors. Appl Phys Lett 89(20):203107-1–203107-3

Pandya ND, Majumder MK, Kaushik BK, Manhas SK (2012) Performance comparison of mixed CNT bundle in global VLSI interconnect. In: Proceedings of IEEE international conference on communication systems and network technologies (CSNT 2012) Rajkot, India, pp 790–793

Pasricha S, Kurdahi FJ, Dutt N (2010) Evaluating carbon nanotube global interconnects for chip multiprocessor applications. IEEE Trans Very Large Scale Integr (VLSI) Syst 18(9):1376–1380

Pu S-N, Yin W-Y, Mao J-F, Liu QH (2009) Crosstalk prediction of single- and double-walled carbon-nanotube (SWCNT/DWCNT) bundle interconnects. IEEE Trans Electron Devices 56(4):560–568

Rabaey JM (2002) Digital integrated circuits, a design perspective, 2nd edn. Prentice-Hall, Englewood Cliffs

Rakheja S, Naeemi A (2010) Interconnects for novel state variables: physical limits and device and circuit implications. IEEE Trans Electron Devices 57(10):2711–2718

Rakheja S, Naeemi A (2012a) Interconnect analysis in spin-torque devices: performance modeling, optimal repeater insertion, and circuit-size limits. In: Proceedings of IEEE 13th international symposium on quality electronic design (ISQED 2013), Santa Clara, CA, USA, pp 283–290

Rakheja S, Kumar V (2012b) Comparison of electrical, optical and plasmonic on-chip interconnects based on delay and energy considerations. In: Proceedings of IEEE 13th international symposium on quality electronic design (ISQED 2012), Santa Clara, CA, USA, pp 732–739

Raychowdhury A, Roy K (2006) Modeling of metallic carbon-nanotube interconnects for circuit simulations and a comparison with Cu interconnects for scaled technologies. IEEE Trans Comput Aided Des Integr Circuits Syst 25(1):58–65

Roslyak O, Gumbs GF, Huang D (2010) Tunable band structure effects on ballistic transport in graphene nanoribbons. Phys Lett A 374:4061–4064

Rossi D, Cazeaux JM, Metra C, Lombardi F (2007) Modeling crosstalk effects in CNT bus architectures. IEEE Trans Nanotechnol 6(2):133–145

Sahoo R, Mishra RR (2009) Simulations of carbon nanotube field effect transistors. Int J Electron Eng Res 1(2):117–125

Salahuddin S, Lundstrom M, Datta S (2005) Transport effects on signal propagation in quantum wires. IEEE Trans Electron Devices 52(8):1734–1742

Sanvito S, Kwon Y-K, Tomanek D, Lambert CJ (2000) Fractional quantum conductance in carbon nanotubes. Physics Rev Lett 84:1974–1977

Sarkar D, Xu C, Li H, Banerjee K (2011) High frequency behavior of graphene based interconnects. IEEE Trans Electron Device 58(3):843–852

Sarto MS, Tamburrano A (2010) Single conductor transmission-line model of multiwall carbon nanotubes. IEEE Trans Nanotechnol 9(1):82–92

Sarto MS, Tamburrano A, Amore MD (2009) New electron-waveguide-based modeling for carbon nanotube interconnects. IEEE Trans Nanotechnol 8(2):214–225

Sathyakam PU, Mallick PS (2011) Transient analysis of mixed carbon nanotube bundle interconnects. IET Electron Lett 27(20):1134–1136

Satio R, Fujita M, Dresselhaus G, Dresselhaus MS (1992) Electronic structure of graphene tubules based on C60. Phys Rev B 46:1804–1811

Shah TK, Pietras BW, Adcock DJ, Malecki HC, Alberding MR (2013) Composites comprising carbon nanotubes on fiber. US Patent, US8585934 B2

Sinha S, Balijepalli A, Cao Y (2009) Compact model of carbon nanotube transistor and interconnect. IEEE Trans Electron Devices 56(10):2232–2242

Srivastava A, Sylvester D, Blaauw D (2005a) Statistical analysis and optimization for VLSI: timing and Power. Springer, New York, p 105

Srivastava A, Xu Y, Sharma AK (2010) Carbon nanotubes for next generation very large scale integration interconnects. J Nanophotonics 4(1):1–26

Srivastava N, Banerjee K (2005c) Performance analysis of carbon nanotube interconnects for VLSI applications. In: Proceedings of IEEE/ACM international conference on computer-aided design (ICCAD 2005), pp 383–390

Srivastava N, Joshi RV, Banerjee K (2005b) Carbon nanotube interconnects: implications for performance, power dissipation and thermal management. In: Proceedings of IEEE electron devices meeting (IEDM 2005), pp 249–252

Srivastava N, Li H, Kreupl F, Banerjee K (2009) On the applicability of single-walled carbon nanotubes as VLSI interconnects. IEEE Trans Nanotechnol 8(4):542–559

Subash S, Kolar J, Chowdhury MH (2013) A new spatially rearranged bundle of mixed carbon nanotube as VLSI interconnection. IEEE Trans Nanotechnol 12(1):3–12

Tsai FF, O'Brien CJ, Petrovi NS, Raki AD (2005) Analysis of optical channel crosstalk for free-space optical interconnects in the presence of higher-order transverse modes. Appl Opt 44(30):6380–6387

Tsukagoshi K, Alphenaar BW, Ago H (1999) Coherent transport of electron spin in a ferromagnetically contacted carbon nanotube. Nature 401(6753):572–574

Wallace PR (1947) The band theory of graphite. Phys Rev Lett 71(9):622–634

Wang N, Tang ZK, Li GD, Chen JS (2000) Materials science: single-walled 4 Å carbon nanotube arrays. Nature 408(6808):50–51

Wang X (2009) Fabrication of ultralong and electrically uniform single-walled carbon nanotubes on clean substrates. Nano Lett 9(9):3137–3141

Wang Z, Zhao G-L (2013) Microwave absorption properties of carbon nanotubes-epoxy composites in a frequency range of 2-20GHz. Open J Compos Mater 3:17–23

Wassel HMG, Dai D, Tiwari M, Valamher JK, Theogarajan L, Dionne J, Chong FT, Sherwood T (2012) Opportunities and challenges of using plasmonic components in nanophotonics architecture. IEEE J Emerg Selected Topics Circuits Syst 2(2):154–168

Wind SJ, Appenzeller J, Martel R, Derycke V, Avouris Ph (2002) Vertical scaling of carbon nanotube field-effect transistors using top gate electrodes. J Appl Phys 80(20):3817–3819

Wei BQ, Vajtai R, Ajayan PM (2001) Reliability and current carrying capacity of carbon nanotubes. Appl Phys Lett 79(8):1172–1174

Wei C, Srivastava D, Cho K (2002) Thermal expansion and diffusion coefficients of carbon nanotube-polymer composites. Nano Lett 2(6):647–650

Wei P, Bao W, Pu Y, Lau CN, Shi J (2009) Anomalous thermoelectric transport of dirac particles in graphene. Phys Rev Lett 102(16):166808

Wilson M, Kannangara K, Smith G, Simmons M, Raguse B (2002) Nanotechnology: Basic science and emerging technologies. CRC Press

Wind SJ, Appenzeller J, Avouris P (2003) Lateral scaling in CN field effect transistors. Phys Rev Lett 91:058 301-1–058 301-4

Wesstrom JJ (1996) Signal propagation in electron waveguides: transmission-line analogies. Matter 54:11484–11491

West PR, Ishii S, Naik GV, Emani NK, Shalaev VM, Boltasseva A (2010) Searching for better plasmonic materials. Laser Photonics Rev 13:1–28

Xu Y, Srivastava A (2009) A model for carbon nanotube interconnects. Int J Circuit Theor Appl 38(6):559–575

Yu MF, Lourie O, Dyer MJ, Moloni K, Kelly TF, Ruoff RS (2000) Strength and breaking mechanism of multiwalled carbon nanotubes under tensile load. Science 287(5453):637–640

Zutic I, Fabian J, Sarma SD (2004) Spintronics: fundamentals and applications. Rev Mod Phys 76(2):323–410